DEAR CHRISTIAN, DO NOT FALL AWAY!

"ABOUT THE TIMES OF THE END, A BODY OF MEN WILL BE RAISED UP WHO WILL TURN THEIR ATTENTION TO THE PROPHECIES, AND INSIST ON THEIR LITERAL INTERPRETATION, IN THE MIDST OF MUCH CLAMOR AND OPPOSITION."

SIR ISAAC NEWTON 1642-1747

MICHAEL B WOELFEL

CONTENTS

STOP...

"Take heed to yourselves, lest...
that day come upon you unawares.
For as a snare shall it come on all them
that dwell on the face of the whole earth"
(Luke 21:34-35).

LOOK...

"But ye, brethren, are not in darkness,
that that day should overtake you as a thief.
Ye are all the children of light, and the children of the day:
we are not of the night, nor of darkness"
(I Thessalonians 5:4-5).

LISTEN...

"Howbeit when he, the Spirit of truth, is come,
he will guide you into all truth ...
he will show you things to come"
(John 16:13).

BELIEVE...

"Man shall not live by bread alone, but by every word
that proceedeth out of the mouth of God."
(Matthew 4:4)

FOREWORD

Dear Christian, ***Do Not Fall Away***, reveals people and activity now in our world, that match Bible passages of the great tribulation. Sobering facts within these pages show the existence of the anti-Christ, today. Yes, Satan will ultimately be defeated and bound (Revelation 20:1-3). Meanwhile, God has given detailed signs for his people to be aware and ready early for a "snare," and the brief reign of the enemy's "beast" minion over earth.

Bible prophecy has been positioned by our Lord lovingly, to identify changes that will challenge people. Foreknowledge written thousands of years earlier gives counsel and affirms that God is personally overseeing and caring for us today. For you to learn about recently activated ancient prophecy, will strengthen your faith and prepare *you* to be a source of relevant useful information, to lead your loved ones during the Lord's closing transition of mankind.

Unlike perhaps anything currently published, Michael's book can serve to assist the Believer to shine as a light in chaos. It is crucial for Christians to embrace the whole Bible-literally, especially at this moment in time, and to grasp what God has clearly stated our role should be in the great tribulation. A relevant question to be answered by each Believer today, 'am I earnestly expecting the literal prophesied alignments and events listed in the Bible, to occur?'

Dear Christian, Do Not Fall Away, will reinforce your worldview with new confidence - verse by biblical verse. Michael begins by offering support for the reader's faith, showing evidence-based scientific findings of earth's biblical origin. Next, you will personally fulfill Bible prophecy with the opportunity commanded in the book of Revelation, to count *"the number of a man."* Lastly, actionable Bible-based ideas are given for the season just ahead.

Christian Philosopher, Blaise Pascal observed: *"Truth is so obscure in these times, and falsehood so established, that, unless we love the truth, we cannot know it"*[1]

Dr. Joel Salazar

INTRODUCTION

Why should we consider the great tribulation at this time? Throughout Scripture, repeated instructions are given for people to stay **watchful** for specific end-of-the-age signs. *"As a snare"* God said, they will come upon *"all"* (Luke 21:35). Right now a 74-year-old Roman Prince is alive, whose name equals 666 in two languages. He also fits each of the many odd features given in the Bible for the anti-Christ. Additionally, in Israel now is a zealous Jewish group that unwittingly desires to perform Old Testament sacrificing on Solomon's ancient temple mount. Here in, ***Dear Christian, Do Not Fall Away***, we will examine how together current people and places appear to align exactly with the biblical drama called the, *"the abomination of desolation."*

Whether these are the actual prophetic performers or not, we will watch and time will tell. Because of the all-encompassing global impact associated with this key event, it certainly must *not* be neglected. Jesus warns us in Mark 13:14,19, that this singular event will be the telltale signal that shows the beginning of the most troubled time in all of human history: *"But when ye shall see the abomination of desolation, spoken of by Daniel the prophet, standing where it ought not, (let him that readeth understand) ... For in those days shall be affliction, such as was not from the beginning of the creation which God created unto this time, neither shall be."*

Much has been written on this subject over the years, but mostly as speculation of what the Bible verses *may* mean. I am a lifelong student of Scripture. Having retired from working as a commercial airline mainte-nance supervisor for twenty-four years, my life's work has been to study, interpret and oversee the application of technical procedural manuals, to cautiously extract the understanding and apply it to a mechanical aircraft - *as if lives depend on it.* You are about to see seamless interconnected facts and living people linked mechani-cally with the Scriptures. It is with trembling awe of God that I submit this work to the sincere scrutiny of fellow believers and anyone seriously seeking truth. I earnestly pray, **Dear Christian, Do Not Fall Away**, will ignite practical dialog, responsive planning and appropriate preparation, for this most trying season in the history of mankind.

May your life and mine be from this moment on, a single-minded endeavor to please our Heavenly Father - who gave us life and put the planet 8000 miles under our feet. May we live for others with first responder zeal, in joyful overflow of undistracted gratitude for the blood that Jesus shed for our salvation! Finally, may you and I live in focused dedication to follow the Holy Spirit, and endure the great tribulation through to our rapture away, or our final natural or sacrificial life's end.

In the due course of history, the height of human value and human purpose became magnificently clarified, as Jesus split time and defeated Satan on behalf of the whole world. He provides the G-O-S-P-E-L, or simply put: God's Own Son Purchased Eternal Life! The pinnacle of our Lord's love is seen in His never-changing pattern today; namely, God provides people with instructions and choices. The Father wants individuals to have open hearts, to accept guidance from His Word, to receive, embrace, and supremely treasure their gift of salvation- His eternal heavenly best! Even so, for the lukewarm, those *not* fixated on Biblical truth, people valuing the million things, above Jesus, and the deceived and deflectors of godly information - an alternative choice exists. It is the sad human default: to *"fall away"* and be eternally doomed in the end.

As a lover of truth, early American founding father, Patrick Henry (1736-1799), shows us the correct posture and determination with which to approach new intimidating information: "For my part, whatever anguish of spirit it might cost, I am willing to know the whole truth; to know the worst, and to provide for it."

Please pray about this paradigm. Only make this your personal prayer:

Heavenly Father, I ask that you cause the message and ideas within these pages to be clear and useful. Please give me discernment to understand your personal objectives for me to have this information. Cause hardy courage to rise in my heart, to respond with abounding fruit of the Holy Spirit in bold faith, and especially with love and compassion for those to whom I am responsible. May I live in accordance with Your Holy Word and as Your Holy Spirit alone directs. Cause me to recognize and cast down all paralyzing fear, and compassionately and diligently serve my family and church as required.

In Jesus' glorious name I ask this, Amen!

"About the times of the end,

a body of men will be raised up

who will turn their attention to the Prophecies,

and insist on their literal interpretation,

in the midst of much clamor and opposition."

— SIR ISAAC NEWTON

1

HUMAN ORIGIN:

OUR ALL IMPORTANT FOUNDATION

"NO!" I shouted at my professor, as he began chalking on the board one morning. He had just announced the day's lecture: "How Man Came from Monkeys." This incident occurred a few months into my freshman college geology class. The professor stopped writing and stood motionless with chalk held against the board. My heart was pounding and felt like it would bounce from my chest. Since the start of this class I was vexed, but could not put my finger on the reason — until that moment. He had just presented a direct affront to my new faith in God and the Bible. After a tense pause he repeated himself, as if pretending not to hear my objection, "Today we will learn how man came from monkeys." With adrenaline now surging, I shouted again even louder, "NO!" He froze a second time, paused, then turned slowly toward the class. With stern calmness he instructed us to open our textbooks to the next chapter, a study of trilobites - the index fossils.

I felt relieved; he would not impose his confusing theory on the class or me. However, as I stood to leave after the day's lesson, it dawned on me, should my professor have responded differently and demanded, "Young man, stand up and give us your account of human origin," I could only have stammered and weakly begun recounting the Genesis, six-day story of naked Adam and Eve, and a talking snake. At that moment I realized a vital need was for me to find if the Bible could somehow be demonstrated as accurate world history. Was it the literal documentation of man's physical past, or was my new belief in God a floating faith – detached from the realities of natural history and science? Was my faith unable to stand up to a rigorous impartial examination?

Questions flooded my mind. Fear swept over me, as I pondered, did a loving Creator have servants write down creation events accurately, or not? Modern science seemed to prove the recent start time given for man in the Bible, to be a myth. These conflicting ideas now ruled my waking hours. Much was at stake: what parts of the Bible can be trusted; did we evolve from animals; and what about dinosaurs? Does life have an overall purpose? I wanted to believe the Bible, but opposing theories taught as facts since childhood, prevented me. With my faith now in limbo, a desperate search to confirm what I believed had begun.

Since every person's important life decisions have been assembled upon some human origin belief, changing this vantage point becomes a more difficult and less desirable undertaking the older we are. What's more, the passing of time tends to harden people's beliefs in general. So, with time and life's decisions long established, the replacement of one's worldview may be the most humbling task a person can embark on. Nevertheless, with eternal life and death at stake, this foundational outlook is perhaps the most worthy understanding for you and I to be absolutely certain that we determine correctly - irrespective of regrets that may result from making a change. Frankly, confessing that one's life has been partially wasted on wrong values, is difficult. One estimate shows less than 5% receive Christ after age thirty.[1]

Just as I left bewildered that day in college, many people today if confronted, are ill prepared to give a sound scientific defense for their biblical beliefs. Stories abound of those who in college abandoned their church involvement. George Barna did the study and found an astonishing 70% of post high school — more particularly, secular college attending Christians, effectively discard their faith by graduation day[2]. This debate is indeed a central but sorely neglected Christian battlefront. I thank God that when I sought, I found answers.

As we begin, let me encourage you to keep a Bible close by. References are plentiful and should be looked up to fully understand the points set forth. Besides, your eternity is way too important to blindly trust it to anyone else than your Creator. Let us start by reaffirming God's declaration of human origin. Chapter one of Genesis launches the definition of human existence and provides the complete origin story of sin and our essential need of faith in the Lord Jesus Christ. Therefore, here *"In the beginning,"* is the natural point for this end-time message to kickoff. In reality, if you and I do not have concrete, indisputable knowledge of exactly where man came from, when, how, and why, then on what basis can we believably explain to others why chaos is going on around us, or even why we exist to begin with?

In the last century, confusion about the origin and purpose of human kind has made individuals and whole societies vulnerable to discarding their priceless faith in God, when the cost to keep it became extreme. Hitler, apparently, understood this fickle nature of people, when he said, "Make the lie big, make it simple,[3] keep saying it, and eventually they will believe it." He observed also, "How fortunate for leaders that men do not think." The Fuehrer boldly promoted his evolutionary based Arian beliefs ("I do not see why man should not be just as cruel as nature."[4]), forcing countrymen to toss their vague worldviews, in place of his willful beastly ideas.

That liar's horrific national influence is well documented. Our world has a biblical ending of global barbarism, as this book will chronicle. Today, similarly, the people most vulnerable to being swept away in the chaos and error, will be those with incomplete knowledge of human origin and God's overall purposes. With this in mind, let us briefly revisit the foundation upon which we each establish our faith, build our lives, and apply our values - mankind's origin.

In 1976, a resurgence of creationism began to take place, much inspired by Dr. Henry Morris' book, *"The Genesis Record"* . The late, Dr. Morris, founder and President Emeritus of the Institute for Creation Research, ICR, is a leading research center in America with a global team of eminent scientists. Among them, according to their article cited below, "57 have earned doctorates in their fields from accredited universities." Hundreds of more scientists around the world, allied with ICR[5], are listed on their website and have similar qualifications.

These notable scientists, with various Christian backgrounds, all mutually embrace the tenets of literal biblical inerrant authority and promote the position that our earth is both young and thoughtfully created. The following biographical information is taken from an ICR *Impact* publication[6], normally devoted to scientific creation evidence. This article reveals the training, competence, and influence of some of the associated scientists.

To show you the reader, the esteemed prominence of the recent-creation worldview, I list a selected four of these Genesis-believing scientists[7]: 1. Kenneth B. Cumming (Dean and Professor of Biology), PhD. from Harvard, where he studied under Ernst Mayr, who is "often considered the dean of living evolutionists"; 2. Dr. Carl B. Fliermans (Microbiology), a microbial ecologist at Dupont with over sixty technical publications. He is well known as the scientist who first identified the "Legionnaire's Disease" bacterium; 3. Dr. Kelly Hollowell (Molecular Biology), PhD. in Molecular and Cellular Pharmacology from the University of Miami. She is also an attorney (J.D.). Dr. Hollowell's work includes a number of publications in the fields of DNA technology, cloning, and neurobiology; 4. Dr. Raymond

V. Damadian, M.D., an inventor, most notably of the M.R.I. Machine.

These, and many other richly credentialed scientists, very intimate with the assorted aspects of the Darwinian macro-evolution theory, have wholly rejected his ideas. There are many scientists today worldwide, numbering in the thousands, who have likewise turned away from the monkey-man conjecture, and now likewise embrace scientifically, the literal Genesis record of human and earth origin - as having begun about 6000 years ago.

We humans are fallible beings, who use fallible equipment to evaluate data from a past that is sketchy and wide open to interpretation. Scientists affirming the theory of evolution, and those embracing Biblical creation, have the same limited capability to evaluate past physical evidence and then test their conclusions. It is, therefore, natural that enormous differences would exist. Both ideas require assumptions be made to fill in large gaps in available data. Such an assessment is similar to glancing at an old man to determine the detailed story of his life.

Because of these large time expanses, all interpretation will naturally have the bias of the one doing the interpreting. Theoretical ideas concerning the earth's origin are after all, on a planetary scale and could never be duplicated, so any final theory cannot be proved. No human alive watched the origin of the earth or changes over lengthy stretches of time. Therefore, all views of man's origin must be based on faith. Christians value living by faith. Secularists despise the concept and mostly define their theories as if they are proven facts. However, as you will see, their ideas are not proven and when examined — most are not even faintly probable.

Although it is encouraging and freeing to discover various physical features that highlight the accuracy of Scripture where it describes this planet, it would be foolish to sit in judgment, attempting to validate Scripture using anything from our surroundings, except Scripture itself. We are finite beings and limited in every way, while God is all-powerful, possesses all knowledge, and is infinite in every way. Before we look at some evidence that supports and refutes today's popular origin beliefs, such as Creation and Evolution, it is useful to understand a major bias during the decades in which the theories have been presented.

Today, we can trace the lopsided dominance of evolutionary views in our culture back not to any testable scientific evidence, but to a single misguided decision of Congress. Evolution centered curriculum and materials flooded into mainstream America in 1959, after President Eisenhower requested and received one billion dollars from Congress for the Department of Health, Education and Welfare, to among many things, publish the old-age ideas. Beginning at the lowest grade levels, an earth history defined by billions of years is now promoted in our schools and media, as fact. In addition, since sparse biblical-creation curricula are put forward even in churches and private schools, the evolution/old-earth ideas continue largely unchallenged.

We Americans now graduate students from high school and college exposed only to the single view of evolution, effectively inoculated against alternate ideas of human origin. The result is a national citizenry that possesses quiet pockets of doubt about the whole Bible's validity. Aggressively teaching just the one side of the issue of human origin has monopolized a free exchange of ideas. At worst, it makes the Bible appear untruthful and God to be either incapable, or else lacking a desire for coherent communication.

Quite clearly, there is no substitute for rock-solid knowledge that supports a person's worldview, particularly if it is our future that becomes threatened, as new evidence cited in the next chapters will point out. Lets face it, if the first pages of Genesis, dealing with the pivotal issues of human origin and human sin are not precise and truthful, then beginning where into the

Bible should we begin to trust it's wording? Are humans, in fact, an advanced link in a billion year old animalistic food chain, or created recently by a fantastic designer? We must know this truth for certain. Let's start with what we can see.

A New View Of Old Things

The enormous cavities of oil and coal buried world-wide, sometimes up to five miles down, and often under what are now the oceans, actually confirm a cataclysmic flood event has occurred in the past. A worldwide flood has quickly buried surface organic matter and vastly changed the earth's crust. As scientist Ken Ham observes, "We see billions of dead things, buried in the rock layers, laid down by water, all over the earth."[8]

Today, apart from man made landfills worldwide, what we don't see are comparably new, naturally occurring, underground organic deposits and reservoirs being formed. Certainly our energy-reliant society at present, would be nowhere near as advanced without trans-portation power and the many products that come from fossil fuels. It seems a source of our current pros-perity, in everything from plastic products around our homes, to the fuel in our cars, trains, and aircraft, even the shingled roofs over our heads, may come from the trapped organic remains of the judged people of Noah's day!

Buried Energy Sustains Us Now

The existence of crude oil and natural gas have in the past been presented as evidence of an old earth age. But what do the buried fossil fuels actually mean? In order for them to form, certain natural forces must be present. First, a rapid sedimentary overlay must occur, as from a flood, landslide, or volcanic aftermath (plants and animals dying on the earth's surface quickly turn to soil, not oil). Massive quantities of organic matter must be covered over and put under pressure, with moderate temperature, over time. Not too much pressure and time however, are needed. The method that was used by the now defunct, *Changing World Technologies*[9], worked with only modest heat and pressure to convert poultry offal and municipal garbage into oil. This process does not take millions of years. Household garbage was changed daily into commercial grade crude oil, in a few hours!

A surprising fact about the twin fuels, oil and coal, is that they are mostly the same, except for water content. Coal can also be formed quickly in a matter of weeks in a lab.[10] Experiments have included using wood chips, compressed at 150 degrees Fahrenheit for thirty-six weeks, resulting in pieces of hardened coal. The Creation Evidence Museum, Glen Rose Texas, displays a skillfully hand-crafted metal pouring cup, embedded in a lump of coal. A miner chiseled it from a solid coal seam, evidence that skilled humans were alive before coal formed.

Science Caves-in to a Young Earth View

Cave formations, once thought to form over vast lengths of time, are being found in man-made structures. One stalactite, over five feet long, was found growing under the Lincoln Memorial that was built in 1923. Thousands of stalactites some over thirty feet long can be seen at the Mount Isa mine in Queensland, Australia. This tunnel was hewn out only sixty years ago.[11] At these rates, all currently discovered underground cave structures can be formed in less than 7,000 years.

Suddenly, Eight Billion People...

Let's continue with more observations of earth's young age, affirming the Bible's account. According to the Encyclopedia Britannica, the current rate of population growth worldwide is 1.7% annually.[12] It doesn't sound like much, but when applied to a large and growing larger world population, the results of compound growth can be amazing. Each year now, the earth's population grows by about 102,000,000 people more than the loss by death. To see this kind of growth in action, we can consider that roughly 330,000,000 people today call the United States of America, home. While just a mere 500 years ago, nearly all land in this country remained untamed and with vast unpopulated areas.

Keeping the above annual growth rate in mind, what should today's population density be if humans had a recent beginning, as in a young-earth origin? Then comparing with the old age paradigm, what would this planet look like if humans originated one million years ago? The results are telling.[13]

Recent or Ancient — What's Reasonable?

First, the young earth scenario. If the population growth rate were charted backward in time from today, at a slower rate of only 0.5% annually, less than one third of today's actual annual rate of 1.7%, the result would be a handful of people alive about 4500 years ago, just as Noah's post-flood world population would have been.

For comparison, on the other hand, what if population growth were at the super slow pace of only .01% annually, an extremely generous 1/170th of today's actual growth? At that snail's pace rate, starting one million years ago, as evolutionists propose, an adventurous monkey and his wife stood up and were considered people, it would mean a doubling of the number of people every 7,000 years, or on average, two increased to four in 7000 years, became eight after 7000 more, etc. A stunning 142 doublings would occur in one million years. That means, at that ultra-slow rate, we should see today, a population amounting to an absurdly large, 10 followed by 43 zeroes of people, or, 100,000,000,000,000,000,000,000,000,000,000,000, 000,000!

To show a contrast to that massive number, consider the earth's total landmass in square meters, is 1.5 x 10 followed by only 14 zeroes. A planet with an impossible population this huge is best understood with an illustration: Earth would have people, side by side, covering the complete surface and stacked one on top of another, up and out in all directions, beyond the distance of our sun! Where are all those human remains? Even accounting for Noah's flood only 4500 years ago (less than one doubling), we look around and see a world today of 7,900,000,000+ people. The rapidly expanding nature of population growth is yet another puzzle piece showing that man was created recently, as depicted in the Bible.

To persist in believing that evolution is true, humanity would have had to remain on the thin edge of extinction over 99% of the last one million years, extremely unrealistic, but worse, an enormous challenge for evolution, which remember by their theory, must have countless random mutations to develop their avant-garde monkey duo. Doubling every 7000 years is hardly conducive to human life, much less for the requirements for the theory of evolution. Again, we see that the recent/sudden biblical origin account stands confirmed by the physical world around us. Until now, we've talked mostly about people, but we live on a planet teeming with a myriad variety of life!

It Is Indeed a Lively World

Life on earth, in its many thousands of varieties and complexities, is alleged to require millions of years to evolve. Yet, hundreds of unfossilized (raw) dinosaur bones, found recently, are not likely more than a few thousand years old, as Dr. Carl Wieland describes: "Evidence of hemoglobin, and the still recognizable shapes of red blood cells, in unfossilized dinosaur bone (T-Rex) is powerful testimony against the whole idea of dinosaurs living millions of years ago. It speaks volumes for the Bible's account of a recent creation."[14]

In another article, after a detailed discussion of the radically different features of one dinosaur from another, the late Dr. Duane Gish concludes his observations: "If evolution were true, we should easily find a series of transitional forms showing these unusual structures gradually coming into being. If creation were true, the dinosaurs bearing these unique structures should appear all at once, fully formed from the start. Here, creation wins hands-down. In not a single case can the required transitional forms be found."[15]

The fact is, the fossil record nowhere supports the theory of evolution, applied to dinosaurs or any animals, living or extinct. The lie of evolution is perhaps the enemy's most subtle and effective tool, to cause people to doubt the Bible, and thus, deflects many thoughtful searching souls from considering any relevance of the Creator in their lives.

A Valid Theory Should Provide at Least Some Evidence

It is common knowledge that the so-called transitional fossils simply do not show progressive similarity. Worse yet, evolutionary theories have never been able to explain what the mechanism is, that causes additional DNA to be created, a result that is necessary if an increase in complexity within something living were ever to occur. Mutations, remember from high school, are mishaps that always result in a scrambling or loss of already existing DNA information, never the required gain of additional organized new information.

It is important to note that no scientist has ever recorded an incident of a living organism evolving in an upward, more complex way. In other words, a mutation has never been reported to have added fresh DNA information, changing a living organism into something more complex, by providing another wing, leg, or an organ. Rather, two wings or two legs, mobility structures oddly 'happen' in balanced pairs.

Today, there is a total lack of evidence for life of any kind evolving from something less complex. Simply put, random Mother Nature has not the creative consciousness to assemble the required additional DNA information for upward changes to occur. The mere passage of time and numberless deaths could not be the extravagant engineer of all life; such an idea is illogical when absolutely no evidence even hints toward that conclusion.

A great deal of scientific technology has been developed since evolution was proposed. Charles Darwin of course, could not have known about these fundamental DNA upgrade limitations to his ideas, when he suggested them. In his defense, Darwin it seems, was at least momentarily intellectually honest. Quoting him from the Origin of Species (chapter VI), "If it could be demonstrated that any complex organ existed which could not possibly have been formed by numerous, successive slight modifications, my theory would absolutely break down."[16] Kudos to Darwin for his honesty!

So What Does Evidence Show Us?

An amazing discovery in the Red Sea, confirms with remains, the miraculous delivering God of Scripture. Similar to a form of Egyptian wall art, pictured below, different coral encrusted ancient chariot wheels, including the crusty wheels and axles, and other artifacts, have been found in the Gulf of Aquaba in the Red Sea. Coral will not form over gold. With the wood structure long dissolved, a single special Egyptian chariot wheel is now only a thin gold veneer on the Red Sea floor (right lower).[17]

> *"The Egyptians pursued, and went in after them to the midst of the sea, even all Pharaoh's horses, his chariots, and his horsemen. And it came to pass, that in the morning watch the LORD looked unto the host of the Egyptians through the pillar of fire and of the cloud, and troubled the host of the Egyptians, And took off their chariot*

wheels, that they drave them heavily: so that the Egyptians said, Let us flee from the face of Israel; for the LORD fighteth for them against the Egyptians."

— EXODUS 14:23-25

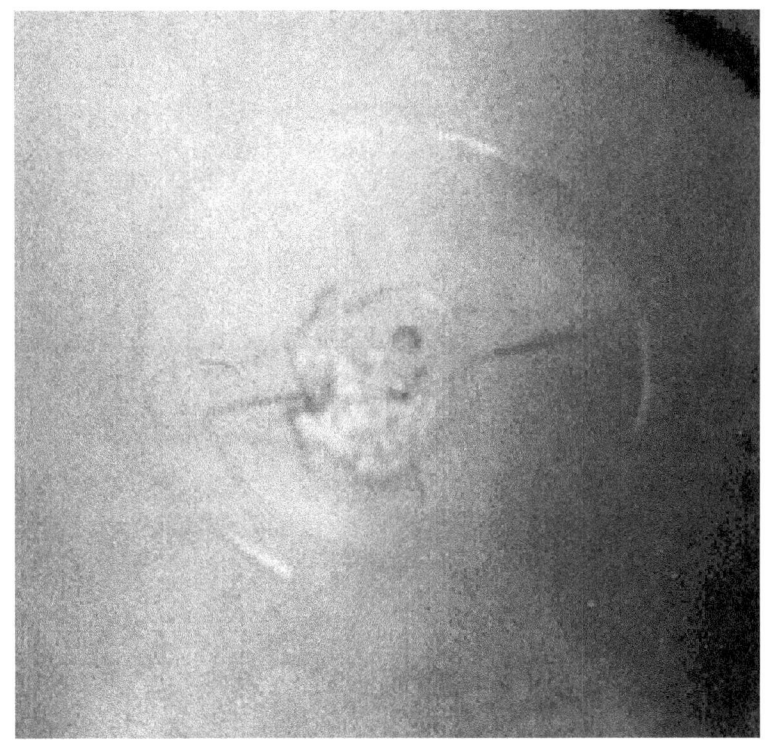

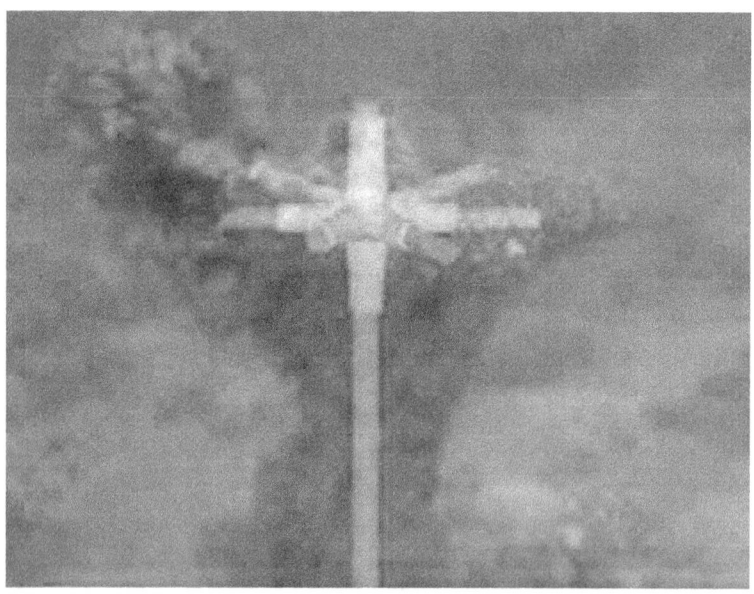

Blind Allegiance Is Not Scientific

There is *no area of scientific study* that does not logically, realistically, and naturally fit within the Biblical model of creation[18]. For decades however, people with dedicated imaginations have interpreted human and monkey bone fragments to be a sub-race called hominids. Assigning them great ages, imaginary monkey-men have been paraded by in magazines and on television. There has been no lack of complete mythical beings, surrounded by virtual family members, displayed in museums around the world.

In his book, "Buried Alive, The Startling Truth About Neanderthal Man," Dr. Jack Cuozzo, a dentist and creationist, has chronicled the results from personally researching and x-raying Neanderthal skulls, found in the back rooms of many of the world's museum archives. He concludes on page 307, "The primary cause of Neanderthalization of the adult skull is age and function within a superior genome capable of extended longevity."[19]

The skulls of Neanderthals are somewhat different, primarily because of the thick upper brow. Some people today have ears and noses that continue to grow with age, and some don't. Likewise, some brow bones of people today keep growing as they age, while some do not. People lived much longer before and immediately after the flood. This allowed more bone growth in certain genetically predisposed people. They were fully human, walked upright, and would scarcely receive a second glance if seen within their first hundred years, on today's streets.

Daily newspapers cooperate to advance the nonscientific dogma, printing multi-million year old claims for any new bone fragments found. Textbooks, from your local kindergarten up to top Ivy League grad-schools, blindly parrot the theory of evolution as fact, while ignoring a whole planet manifest with intricate biological design. The misguided final fallout is sadly worldwide, as secular people eventually develop God-calluses

- a hardened view against consideration of a loving Creator's personal involvement in their affairs.

Nonetheless, the following five verses of Scripture substantiate that God's creative words, spoken in Genesis 1, announcing, "Let there be..." to bring everything into existence, are in fact, the very words now sustaining and holding all matter in this universe together: "*[Jesus Christ] who being the brightness of his glory, and the express image of His person, and upholding all things by the word of His power.*" (Hebrews 1:3); "*For in Him we live, and move, and have our being*" (Acts 17:28); "*The heavens and the earth, which are now, by the same word [at creation] are kept in store, reserved unto fire against the day of judgment and perdition of ungodly men*" (II Peter 3:7). Also, "*if he [God] gather unto himself his spirit and his breath; all flesh shall perish together, and man shall turn again unto dust*" (Job 34:14-15). And lastly, "*For by him [Jesus Christ] were all things created, that are in heaven, and that are in earth, visible and invisible, whether they be thrones, or dominions, or principalities, or powers: all things were created by him, and for him: and he is before all things, and by Him all things consist [hold together]*" (Colossians 1:16-17).

Certainly, we all enjoy the amazing scientific advances of this modern day; a testimony to man's creativity and ingenuity. We are warned by the Apostle Paul however, not to trip on *"oppositions of science falsely so called"* (I Timothy 6:20). A whole lot has been learned that defies Darwin's faith-based antique ideas. Yet the old-age/evolutionary theories are still permitted to lurk among the authentic sciences, misleading people away from wholehearted trust in the saving truths of Scripture. More and more however, thoughtful people are coming to recognize the Creator's obvious intelligent design, a design that includes and encompasses everything we see and know. It is written: *"Through faith we understand that the worlds were framed by the word of God, so that the things which are seen were not made by things which do appear"* (Hebrews 11:3). It could not be said any more clearly. (Also, see Psalm 19:1-2 and Romans 1:20).

Your Opportunity of a Lifetime, and Eternity Beyond!

Finally, dear reader, even if you could win the lottery every day for the rest of your life, it would be a poor exchange for the permanent joy of eternal life given to everyone who receives Jesus Christ as his or her Savior. Since no person has asked to be born, God - true to his character, has simplified the solution to the sin problem that we all have. In fact, our life is a little like a pop quiz, with one pass or fail question... to which God gives the answer! He longs for every person to make the simple right choice. It is written: *"I [God] have set before you life and death, blessing and cursing: therefore **choose life** that both thou and thy seed may live"* (Deuteronomy 30:19).

All of the sins that you and I have committed have been paid for!

The following Scriptures are offered to make this clear:

> *"Jesus saith unto him, I am the way, the truth, and the life: no man cometh unto the Father but by me"*
>
> — (JOHN 14:6)

> *"These are written, that ye might believe that Jesus is the Christ, the Son of God; and that believing ye might have life through his name"*
>
> — (JOHN 20:31)

> *"He that believeth on the Son hath everlasting life: and he that believeth not the Son shall not see life; but the wrath of God abideth on him"*
>
> — (JOHN 3:36)

"For God so loved the world, that he gave [allowed to be a substitute sacrifice] His only begotten Son, that whosoever believeth in Him should not perish, but have everlasting life"

— (JOHN 3:16)

" behold, now is the day of salvation" (II Corinthians 6:2);
"Believe on the Lord Jesus Christ, and thou shalt be saved, and thy house"

— (ACTS 16:31)

You may be saved right now! Pray something like this, but from **your** heart: God be merciful to me, a sinner. I believe Christ died for me and that His precious blood will cleanse me from all my sin. By faith, I now receive the Lord Jesus Christ into my life as my Lord and my Savior, trusting Him for the salvation of my soul. Help me Lord, to know and do your will each day. In Jesus' name, I pray, Amen!

As a new believer, share your faith: *"That if thou shalt **confess with thy mouth the Lord Jesus**, and shalt believe in thine heart that God hath raised him from the dead, thou shalt be saved."* Friend, if you prayed that prayer, and committed your life to Christ, these next verses of Scripture now apply to you:

- *"Therefore being justified by faith, we have peace with God through our Lord Jesus Christ"* (Romans 5:1);
- *"Peace I leave with you, my peace I give unto you: not as the world giveth, give I unto you. Let not your heart be troubled, neither let it be afraid"* (John 14:27);
- *"And the peace of God, which passeth all understanding, shall keep your hearts and minds through Christ Jesus"* (Philippians 4:7);
- *"Eye hath not seen, nor ear heard, neither have entered into the heart of man, the things which God hath prepared for them that love Him"* (I Corinthians 2:9).

Read your Bible daily, for it is the primary food for your spirit, and your new life is nourished by it: *"Faith cometh by hearing, and hearing by the word of God"* (Romans 10:17). Talk with God moment by moment, as you would to a friend. He loves you and desires a personal relationship with you; that's why He made you and me!

Finally, find a church where the Bible alone is believed and preached as the Holy Word of God. *"In every thing by prayer and supplication with thanksgiving let your requests be made known unto God. And the peace of God, which passeth all understanding, shall keep your hearts and minds through Christ Jesus"* (Philippians 4:6-7). If you prayed and gave your heart to the Lord Jesus, Congratulations! Heaven is your new eternal home no matter what you experience in the months to come!

God Is Indeed Communicating Clearly

As we have seen, the differences between creation and evolution are in stark contrast to one another. The heart of the clash actually centers on two factors. The first is the question of *when* in the past, death began to occur on earth. According to the Genesis account, the onset of everything dying came **after** Adam sinned. Examining the biblical creation events, we see that God defined each creation day numerically, a first through a seventh day. Obviously, being God, he could have made all creation in an instant, but chose instead to take seven days and establish a rhythmic pattern - the week, for time in the future.

The Lord is neither clumsy nor deceptive with His communication, but quite the contrary. He is devoted to us having an accurate understanding of Himself and His works. So, as if to prevent any hint of vagueness, the Creator standardized His creation days, by repre-senting each day with an evening and a morning (Genesis chapter 1). Repeating each day, the term "evening and morning," reinforces our normal understanding of an ordinary day. Assuredly, all vegetation made on day three would have had no problem waiting for the sun to be created, within twenty-four hours, on day four. In addition, a standard day's length would most certainly have been used, on days six and seven, as the Lord created and then introduced Adam, who was made on day six, around his new garden home. Ending the week, Adam enjoyed the seventh day of rest with his Creator, not a lazy thousand, or millions of years. On the other hand, to claim these rotating-earth, "evening and morning" kind of days each contain millions of years, would characterize the Creator as having presented to us a strange, confusing, and repeated deception. God cannot lie, and neither would He ever deceive us. Finally, to say these creation days were anything other than standard twenty-four hour periods is to distort the plain reading of the verses into a private interpretation, something Scripture speaks sternly against. (See II Peter 1:20).

Dear reader, Almighty God will not be stammering in response to people who stand to challenge Him on judgment day, claiming to have been misdirected by inaccurate wording in His "misleading" Scriptures. His readiness to righteously judge all people of this world pivots upon the most extreme cost to Himself, the very life of His righteous Son. We can rest assured that His qualification to judge every individual is, and will be, altogether irreproachable.

God Means What He Says

Words have intended meanings. We see that our loving Creator went to great lengths to include clearly written criterion, apparently in anticipation of the coming worldwide deception of millions of prior years of evolutionary death. Again, the Bible teaches that the introduction of death to this earth came specifically because of Adam's sin, sometime after day six. Otherwise, earlier in the six-day creation week, as God pronounced everything "good" and "very good," things would have been created dying and rotting, as part of His original sinless design! What a confusing concept, that clashes with our knowledge of God, and undercuts his carefully defined historical chronology.

In Romans 5:12 we read, *"Wherefore, as by one man sin entered into the world, and death by sin; and so death passed upon all men, for that all have sinned."* And not just death for men: *"For we know that the whole creation groaneth and travaileth in pain together until now."* (Romans 8:22). To tout a new creation made with death and decay present, or one that formed slowly over millions of years is not justified in the Bible, but is actually a doubt-provoking untruth, against our plain spoken Creator. Furthermore, with everything today reproducing *"after its kind,"* all theories of gradual random progressive development, that claim millions of years of mutations and death, plainly do not fit any current observations of life on earth or past remains.

Facts From a Seamless Lineage

The second part of the creation/evolution conflict clashes on the age of the earth, as we have discussed. Jesus, Himself, restates a very simplistic understanding of creation, in Mark 10, verse 6, *"from the beginning of the creation God made them male and female."* The obvious intended meaning of the words in Genesis, and Jesus' restatement of them here in Mark is, well, simple. At the earliest moments of God's creation, man was made - period.

We see here that Jesus accepts the straightforward Old Testament origin account of the universe and man. This verse in Mark may therefore, act as a kind of contextual plumb bob, aligned for us by Jesus. In using a single isolated verse, straight down from the Genesis creation account, our Savior teaches us to accept and refer to the plain meaning of the wording - that of a simple *literal* reading. The creation description is not metaphorical, or given an obscure, ultra-expanded, or private meaning, even by Jesus.

The Bible goes even further to secure its time referencing integrity. We are given a record of the unbroken lineage of names of every father with his son, in succession, with their related ages. Starting with Adam, the Bible seamlessly shows succession to Jesus: Adam goes to Shem, in Genesis 5:3-32, then from Shem to Abraham, in 11:10-26. Lastly, Abraham goes to Jesus, in Matthew 1:2-16. If still in doubt, the lineage is given again backwards, from Jesus to Adam, in Luke 3:23-38. God has given conclusive time-documentation, *"from the beginning of the creation* [and the first man, Mark 10:6]," to the birth of Jesus. The total age of the earth is the simple sum of that lineage, equaling around 4000 years, plus about 2000 years since Jesus walked the earth, up to today.

Creation is about 6000 years old. So our loving Lord, in wise foresight, has made specific provision to prevent any room for billions of years of the alleged death and destruction that evolution is presumed to require. Claims otherwise are false. With the origins battle line fixed, our challenge is to embrace and communicate a contrary worldview. When someone came to Jesus, inquiring what work he might accomplish for God, he was told simply to *"believe on him whom he hath sent"* (John 6:29).

A Sinless Recent Creation is the Bedrock of Christianity

It was by one man, Adam, that sin entered the world; with sin came death. Consequently, we understand that it is from Adam's original sin whence the need for salvation originated. That first sin (reemphasized by every person since) initiated the necessity for the sacrificial death of the sinless Son of God. This is the basis for the Gospel. The original entry time of sin, followed with death's entry, are certainly not points that Christians may dispute. To contradict that the first man, Adam, was created sinless on an earth where death was nonexistent, is to undermine the very reason Jesus gave His life on the cross. It was to conquer sin and death... brought about by a fully developed, thoughtful man's rebellious choice!

Today, many people live bound by the popularized claims of prior mega-years of death and destruction, even in the church, not realizing that those beliefs have accompanying doubts that hinder wholehearted acceptance of the whole Bible as God's inerrant Word. The undermining doubts that come from accepting some form of evolution, or the equally conflicting idea of a progressive creation, have had an impact on the faith of Christians around the world. To see advanced results of that thinking, we need only look to England where modern evolution was birthed. There, church attendance has dropped by over half from 1980 to 2000. While today, less than ten percent of their population now finds any church relevant enough to attend[20].

An Early Preacher Made the: Creation Foundation Central to the Gospel.

How important is biblical creation understanding? We see in Acts, an interesting group of folks who wanted all their bases covered. They had made a space for an *"unknown god."* Notice that Paul introduced those clueless to their Creator by showing off the Lord's creative works, thus laying a solid creation foundation. Let's look at the clever tact the Evangelist used, so the Gospel (with creation verses highlighted) would be relevant to a particular people group in history:

> *"Then Paul stood in the midst of Mars' hill,*
> *and said, Ye men of Athens, I perceive that*
> *in all things ye are too superstitious. For as*
> *I passed by, and beheld your devotions, I*
> *found an altar with this inscription, TO*
> *THE UNKNOWN GOD. Whom therefore*

ye ignorantly worship, him declare I unto you. **God that made the world and all things therein, seeing that he is Lord of heaven and earth, dwelleth not in temples made with hands; neither is worshipped with men's hands, as though he needed any thing, seeing he giveth to all life, and breath, and all things; and hath made of one blood all nations of men for to dwell on all the face of the earth, and hath determined the times before appointed, and the bounds of their habitation; that they would seek the Lord, if haply they might feel after him, and find him, though he be not far from every one of us: for in him we live, and move, and have our being; as certain also of your own poets have said, For we are also his offspring. Forasmuch then as we are the offspring of God,** *we ought not to think that the Godhead is like unto gold, or silver, or stone, graven by art and man's device. And the times of this ignorance God winked at; but now commandeth all men every where to repent: because he hath appointed a day, in the which he will judge the world in*

*righteousness by that man whom he hath
ordained; whereof he hath given assurance
unto all men, in that he hath raised him
from the dead."*

— ACTS 17:22-31

Paul considered the message of creation central when introducing Christianity. Notice, if you read this excerpt in Acts 17 above, without the highlighted creation verses, the Gospel appears baseless and flimsy. We see, instead, that the Apostle Paul took advantage of their empty place erected for an *"unknown god,"* and filled it in for them.

Creation evangelism was necessary for Paul's listeners, and is as much in need now. The scenario today is somewhat different, in that many credit inane theories, or an invisible 'Mother Nature,' not idols as in Athens of old, as the cause for the universe to exist. The piercing light of God's Word is of course effective, but only if people accept the Bible as truthful and relevant.

Encouraging a sinning friend, neighbor, or relative to turn, accept Christ, and *"go and sin no more,"* may conclude fruitlessly, only because the person has his whole life positioned on the wrong foundation of origin. To change his view, he may simply need to learn evidence that the sky he lives under and the planet he stands on, are actual displays of the stunning recent works of a supremely loving Creator, whose documented verbal commands made and sustain all things. With that realization comes fresh clarity, that the rest of the Creator's written statements probably matter very much, too. *"Let God Be True, but Every Man a Liar"* (Romans 3:4).

To summarize, the Lord is not afraid of the results if all scientists were to unite, sharpen their pencils, and begin busily working around the clock to disprove Him. He created math, physics, and all scientific disciplines to bring out His orderliness and glory. Today, an array of technology has been developed and focused to predict daily weather, for the sensitive needs of crop management and commercial air transportation, yet no one can determine if it will snow or be warm three weeks in the future. I'm not being critical, but simply making the point that the accuracy of hindsight technology, like forecasting, is limited by all unverifiable data that must be involved. The wonder and awe humans have always had, discovering God's grand creation, were most certainly anticipated by our Lord when He told mankind to take dominion of a planet with such extravagant intricacy and mind-boggling inventive variety.

The stars were easy for God to create, and while a most significant part of all creation, they have perhaps the least influence in our lives. God categorizes them to us like an afterthought: *"he made the stars also"* Genesis 1:16, (Search: Scoopwhoop, NASA, largest picture ever taken. Watch the stunning 3 minute video).

You, however, were purchased with the priceless life of God *Himself*, setting your and my individual value higher than a mountain of diamonds. He wants you to know that He even has constant count of the number of the hairs on your head, updated since your last shower! (See Matthew 10:30.) Certainly this world is thriving today from the many benefits developed from plain applications of modern science. The Bible however, must always and in every way be considered truthful, while any man, theory, denomination, or idea, conflicting with Scripture, should be viewed decisively as false.

A Concluding Look at Our: Magnificent Father's Wholehearted Love

A loving Creator has provided biblical writings that reliably represent mankind's origin and our current purpose. We can rest assured that this foundation is dependable, and also that the King of the book has provided an accurately prophesied future! The Lord has multitudes of righteous angels, and a third as many ungodly, studying all His judgments. It follows that the Father, by giving His Son as the ultimate sacrifice for man's vile sins, to satisfy a severe and uncompromising heavenly justice/mercy balance, has demonstrated limitless love, faultless character, and has proven He is completely trustworthy, even facing death — both His own in the Person of Christ, and someday ours. Such an extreme offer of love, to defeat the effects of your and my ghastly sin, demonstrates that God is unquestionably passionate that His own communication be accurate, and dear friend, **that He is believed**.

The lies of evolution counter what the Lord has said and are now deflecting many millions of *'educated'* people, around the world, from the saving truth of the Gospel. Moreover, Christians today, who have not embraced the reliable and precise Genesis account, do not live with their roots of faith gripping very deeply into the Bible. Nor are they likely to have absolute confidence in the Lord's loving character, as time concludes. Perhaps most importantly, the rock solid foundation of biblical creation can be an anchor for us, should we be the people who enter what Jesus described, will be great tribulation. Avoiding the mark of the beast, at any necessary cost, is less challenging knowing that our priceless faith like roots, has grown down through the shifty topsoil of human ideas and events, and now grips firmly into the crevices of original bedrock - God's entire infallible Word. The battle for eternal human souls has many fronts, using creation information to both evangelize seekers and fortify today's believers has been too long neglected.

Please share this information. The whole Bible must be embraced for wholehearted faith, beginning with the very first verses! Next, in Chapter 2, we will change our focus completely. From the start of man, we will look into the biblical finish. We will begin with a fashionable idea, widespread among believers. These thoughts carry enormous weight in Christian circles today and must be addressed head-on with strict faithfulness to Scripture. Will people be raptured before, or after the great tribulation? Let's look carefully together and unwrap what the Bible alone has to say on this influential topic.

2

THE ULTIMATE FINISH LINE

The rigid religious leaders of Jesus' day stumbled in the transition God had them to face. Being stubborn, self-righteous, and power-hungry, they rejected Jesus and chose to cling blindly to the status quo. We can learn from them, especially to avoid placing strict requirements for how God must fulfill His unfolding prophecy, in order for us to be receptive. Perhaps God has planned to open the verses revealing Jesus' Second Coming the way He did for His first coming: to be confirming guidance that enables the people approaching Jesus' transitional return, to orient themselves using prophetic verses, **as** events begin happening. Yet, even though they had large portions of the Old Testament memorized, those stubborn religious Pharisees missed the miracle-working Savior in their face, rebuking them.

God's appointed leaders in Jesus' day sadly failed their test when challenged to personally re-evaluate their doctrines. Looking back, we learn that a dismissive, proud, or doctrinally rigid leader can be destructive to those he influences. Nevertheless, our Lord repeatedly expresses in the Bible, that end-time passages are to be prayerfully evaluated. While the luxury of cherry-picking certain doctrines and ignoring related more extreme passages, to maintain comfort, security, or a position, has never been an option permitted in the Bible. I have wrestled, prayed, and wept occasionally at the overwhelming meanings, while writing this book. I cannot face my Creator without having made these discoveries available. What foot soldier, watching on the wall, is not expected to sound an alarm, if he sees telltale enemy movement... *especially when a battle has been foretold and is expected?*

After 2,000 years of the Lord's stable gentle grace, we must shake ourselves alert to examine the possibility that actual prophesied events may now be in our faces. But let us instead, humbly recognize the realistic possibility, that events again may be unfolding very differently than the current majority have been expecting. First, let us stir ourselves with a review of some historical changes that occurred to people in the Bible, and examine how they handled them.

As we look back in a moment, we see that the Lord has not demonstrated a pattern of providing our forerunners with perfectly clear details, before they faced their times of change. However, the Lord has shown a consistent pattern of allowing total freedom of choice. People who responded contrarily, were allowed to transition away from opportunity into surprising trials and even destruction. The nature of the tests and trails, in every case, were to qualify them. Challenges were to determine whether people would respect or disregard - *God's written or spoken word* – in favor of their perception, surroundings, peers, enemies, temptations, or physical situation. God wanted to see if they with focused determination, would believe and obey Him. Tests for man through the ages it seems, were really simple... for six thousand years! As you look at the following brief overview, you will notice each opportunity people were presented, included a promise or instruction from God, along with options... one of course, was to faithfully trust and obey Him:

Adam and Eve transitioned from bliss to curse in a surprising way; they failed, for all mankind, a simple pivotal test of obedience. Later, the whole world became overwhelmed after Noah tried but couldn't convince anyone, outside his immediate family, of a coming watery transition.

Also, Abraham's descendants had for generations, been the chosen diligent caretakers of the Old Testament Messianic prophecies. Yet, most selfishly rejected all transition that accompanied the Messiah when He finally came. God's stunning response was to temporarily reject His people and transition two thousand years of unclean Gentiles in!

The Lord used Moses to transition the oppressed Israelis to freedom from Egyptian slavery, through one miracle after another. Astonishingly, all but two men of the millions who transitioned out of bondage, never transitioned into the promised land. They murmured and failed a faith-in-God/fear-of-man test.

Jesus provides the vital, long awaited and exclusively celebrated, everlasting transitional opportunity, for all mankind. He offers salvation and swings wide eternal access to God, but generally only common people accept Him. Even Jesus' personally mentored disciples, possessing special direct knowledge that their Lord would transition through death and rise again, all scattered surprised and confused, and until they saw Him alive, were despondent.

As a general assessment, very few people through the ages, have embraced what God says, though He makes clear announcements and gives distinct key instructions. Do you see an overall outrageous pattern of human failure, simply listening to and responding in faith to God's plain spoken directions, particularly at transition points, even though He is perfectly clear?

So, here we are, with a need now for sensitivity to and awareness of God's next and final transition of the human race; a coming period Jesus describes, when there shall be affliction, *"such as was not from the beginning of the creation which God created unto this time, neither shall be"* (Mark 13:19). Notice first, that the Lord warns us of the sudden nature of its arrival. Look again at His warning:

> *"Take heed to yourselves, lest at any time your hearts be overcharged with surfeiting, and drunkenness, and cares of this life, and so that day come upon you unawares. For as a snare [snare: to bring into unexpected evil, perplexity, or danger] shall it come on* **all them that dwell on the face of the whole earth"**
>
> — (LUKE 21:34-35)

It will not please our Father if we disregard His Word when we face this scheduled chaotic time. How sad if we end up surprised, doubtful, and leading our family or church plummeting into confusion. Let us learn from our many forerunners, who with clinch-fisted expectation and rigid self-serving doctrine, tripped and failed, many in the name of God. May we be cautious to avoid flattering ourselves, reasoning that our God of Grace would not allow into our lives worldwide transition and upheaval. Will this be our chance to transition into rewards, or if wrongly responded to, into a surprising time of testing, even destruction?

A Perplexing New Source of Doctrine: "Christian Fiction"

Just as our biblical understanding of mankind's origin and fall forms the basis of our Christian worldview, and greatly affects the growth and durability of our faith, what we believe about our future can have even greater influence in the plans and decisions we make today. Naturally, everyone has a desire to know what the future holds, yet the Word of God is complete and no new prophetic Bible wording is being added. Complicating our ability to discern the future, the Lord can exercise an option that has mostly not been acknowledged. Namely, God may keep prophetic understanding hidden for millennia, as He chooses. He embeds meanings in parables, symbolism, prophecies, perhaps future events, and in a host of other ways, that may be understood only when His revealing time is ready. The overarching doctrinal push consistent in most all of today's prophetic scenarios, is the greatly preferred message that Christians will be removed from the earth before the great tribulation begins.

"Study to Show Thyself Approved unto God" (II Timothy 2:15)

Instead of the rapture happening before the tribulation, many mature believers today, understand from their Bible study, that the great tribulation will occur first. In that case, might we hear unyielding and weak believers prophetically scoff, *"Where is the promise of his coming?"* (II Peter 3:4). Think about this hopeless quote; it would only relate to expectant followers, who were anticipating a time and/or sequence for Jesus' second coming, but who then found themselves facing traumatic events so convincing, that their doctrine is proven to be mistaken- that they blurt out the confused verse.

Expressing such a disgruntled phrase may indicate these could be the disillusioned Christians foretold, who would end their faith by *"falling away."* A heartbreaking time for believers; it is an event headlined 2000 years ago. This exodus *"away"* will certainly be a polarizing event in Christendom. The Lord names these defecting people *"scoffers."* Take special notice, they are identified by their vocal skewed view of history; specifically, inaccurate biblical *earth* history, as shown in these verses: *"Knowing this first, that there shall come in the last days scoffers, walking after their own lusts, and saying, Where is the promise of his coming? For since the fathers fell asleep, all things continue as they were from the beginning of the creation. For this they willingly are ignorant of, that by the word of God the heavens were of old, and the earth standing out of the water and in the water: whereby the world that then was, being overflowed with water, perished: but the heavens and the earth, which are now, by the same word are kept in store, reserved unto fire against the day of judgment and perdition of ungodly men"* (II Peter 3:3-7).

This passage alerts us that in the last days there will be lustful cynical people who have fixed ideas about the end-times, so much so, they stubbornly scoff at God's unfolding order of the Second Coming of Jesus! Though they do refer to some form of creation, their fundamental flaw is that they are determined not to accept the literal, physical earth history events given in Genesis. Notice that they have made the fatal mistake of *"willingly"* choosing to be *"ignorant"* that from the word of God, the heavens, the earth, and the Noahic flood are given their definitions.

Naturally, their disregard of "*the same word*" will leave them baffled over the final fiery judgment. From these verses, it is clear that some creation believing people, who do not embrace the Bible's wording of the earth's origin and flood verses, will lack God's foundation and will be among those who scoff at the unfolding order of the end times.

Attention, Christian Soldiers!

Should we trust that our righteous Holy Lord is now advocating fictional Christian book drama and apocalyptic movie entertainment, to inform and prepare this generation with His warnings of severe global tribulation judgment? Wouldn't a choice by God to endorse the assistance of fiction and fantasy to communicate end-time doctrine operate counter to the purity of His divine character?

In contrast to an imagination-based approach to interpreting Scripture, look at the tone of concern our righteous Lord has that we understand it is He alone, and not men, who originated and gives the interpretive understanding of prophecy: "*Knowing this first, that no prophecy of the scripture is of any private interpretation. For the prophecy came not in old time by the will of man: but holy men of God spake as they were moved by the Holy Ghost*" (II Peter 1:20-21). Continuing on into the next chapter, verses 1-3, the Lord of course knew what doctrinal dangers these times would present. He warns us of *"false teachers"* bringing *"damnable heresies"* who use *"feigned* [false or fictitious] *words"* at the end of time (all of II Peter, chapters 2 & 3, describe deceptions of the end times).

Unlike those original wise men bearing gifts, who presumably scrutinized Old Testament documents and studied heavenly signs in search of the first coming of the Messiah, are we correct to think a couple of hours with a work of fiction, or a movie and a bucket of popcorn, apprises contemporary wise men of the Second Coming? One might argue, 'But surely the Lord wants such information broadcast by every means available, doesn't He?' No one can doubt that Almighty God has permitted the pre-tribulation rapture doctrine to become popular. Fascination with the number 666, and current 'end time fervor,' has probably peaked because of the preferable early-exit idea. Additionally, most Christians today live believing such discomfort is for others; the heathen and Jews must deal with it.

But what if, on the other hand, the pre-tribulation position is an allowed error, and Believers alive at the time, will be among "*all*," as the verse says, who are snared as the tribulation begins? Then it would mean, to avoid the mark of the beast, everyone alive must choose (or reject) God afresh, as the tribulation starts. Weak Christians and those rigid in doctrine and unwilling to adjust might be provoked to scoff and be tempted to fall away.

A major recommitment would certainly be necessary as recognition of the great tribulation begins - by all believers, and too, a resolute decision required of those caught up in the rapture-tainment. It would mean that God is mercifully giving a final line-in-the-sand, worldwide call for earth's last living people to choose salvation, sealing their eternal decision for heaven. All frightful who turn and take the mark, to buy and sell in Satan's emerging economy, will seal their choice of hell's fire (see Revelation 14:9-11). Every person equally, would have to persist in rejecting the mark of the beast, and cling to salvation with resolute faith in the death and resurrection of the Lord Jesus Christ.

We go now to the heart of this issue: What is written in your Bible? Aside from what is popular today, there are many who understand the Bible to declare, Christians will endure the tribulation judgment period, and be caught up before the "*wrath*" of God finalizes the earth as we know it. After twenty-eight years with the former doctrine, closer examination of Scripture has changed my understanding.

Perhaps, you will do better than I and numerous others have, in finding error in the following article, rewritten here word for word as stipulated by the publisher. *"The Rapture When?"* authored by Jeff Mott, is included below. This may be the plainest presentation available of the biblical sequence of our Lord's end-time events. Will God's "catching away" occur before the worldwide tribulation trauma begins - as has been widely advertised? I sincerely wish I could say yes, but your Holy Bible does not appear to order it that way.

The Rapture When? By Jeff Mott[1]

What Do the Scriptures Really Say About the Rapture? To get an understanding that comes from God's Word and not from our own expectations of God's plan, we need to see what each scriptural reference is really talking about. We also need to establish what events take place for each verse and when they occur.

Let's start with the favorite verse quoted to support the rapture of the church at the start of the tribulation.

> *"For the Lord himself shall descend from*
> *heaven with a shout, with the voice of the*

archangel, and with the trump of God: and the dead in Christ shall rise first: Then we which are alive and remain shall be caught up together with them in the clouds, to meet the Lord in the air: and so shall we ever be with the Lord. Wherefore comfort one another with these words"

— (I THESSALONIANS 4:16-18, KJV).

Sequence of Events:

1. The Lord descends.
2. A shout, voice, trump.
3. The dead rise.
4. The living caught up.

Timing: Nothing specifies a particular time for these events. In this verse we found nothing conclusive to be used to identify the specific time when the rapture might occur except that it follows the Lord's descent from heaven and the sounding of the trumpet of God.

Let's take a look at another very popular verse about the rapture of the church:

"Behold, I shew you a mystery; We shall not all sleep, but we shall all be changed, in a moment, in the twinkling of an eye,

at the last trump: for the trumpet shall sound, and the dead shall be raised incorruptible, and we shall be changed" (I Corinthians 15:51,52, KJV). Sequence of Events:

1. The trumpet sounds.
2. The dead rise.
3. The living changed.

Timing: Begins with the "last trump". The trump identified as the "last trump" seems to be the cause of these events. This appears to indicate that other trumpets will sound before this one, but none after it. Could this be the final (seventh) trumpet found in Revelation? If so, that would put it at or near the end of the tribulation. Wait a minute! This isn't what is being taught about the rapture of the church. Are there any other verses that can help clarify this matter?

Yes! In fact, Jesus himself explains to His disciples what to expect in the end times. There are two sets of verses regarding Jesus explanation:

> *"Immediately after the tribulation of those*
> *days shall the sun be darkened, and the*
> *moon shall not give her light, and the stars*
> *shall fall from heaven, and the powers of*
> *the heavens shall be shaken: And then*
> *shall appear the sign of the Son of man in*

heaven: and then shall all the tribes of the earth mourn, and they shall see the Son of man coming in the clouds of heaven with power and great glory. And he shall send his angels with a great sound of a trumpet, and they shall gather together his elect from the four winds, from one end of heaven to the other"

— (MATTHEW 24:29-31, KJV)

Sequence of Events:

1. The Tribulation ends.
2. Sun, moon darkened — stars fall.
3. The sign of the Son of man seen — Son of man coming.
4. Angels sent with sound of trumpet.
5. Elect gathered from four winds (earth) and one end of heaven to the other.

Timing: The coming of the Son of man and his gathering of the elect follows the darkness at the end of the tribulation.

And a second set of verses covering Christ's description of the end times:

> *"But in those days, after that tribulation, the sun shall be darkened, and the moon shall not give her light, And the stars of heaven shall fall, and the powers that are in heaven shall be shaken. And then shall they see the Son of man coming in the clouds with great power and glory. And then shall he send his angels, and shall gather together his elect from the four winds, from the uttermost part of the earth to the uttermost part of heaven"*

> — (MARK 13:24-27, KJV)

Sequence of Events:

1. The tribulation ends.
2. Sun, moon darkened — stars fall.
3. The Son of man seen coming.
4. Angels sent.
5. Elect gathered from the four winds (earth) and the uttermost part of heaven.

Timing: Again, the Son of man's coming and the gathering of his elect follow the darkness at the end of the tribulation.

In both these references we see it clearly states that the tribulation will end with the sun and moon darkened and the stars falling. After all the people of the world see the sign of the Son of man, he causes the sounding of the trumpet and sends his angels to gather both the dead (in heaven) and the living (on earth) saints. There should be little doubt left here, that Jesus says these events occur after the tribulation.

Are there any other scriptures that can either confirm or refute this conclusion?

Yes.

> "And the seventh angel sounded; and there were great voices in heaven, saying, The kingdoms of this world are become the kingdoms of our Lord, and of his Christ; and he shall reign forever and ever. And the four and twenty elders, which sat before God on their seats, fell upon their faces, and worshiped God, Saying, We give thee thanks, O Lord God Almighty, which art, and wast, and art to come; because thou hast taken to thee thy great power,

*and hast reigned And the nations were
angry, and thy wrath is come, and the time
of the dead, that they should be judged, and
that thou shouldest give reward unto thy
servants the prophets, and to the saints,
and them that fear thy name, small and
great; and shouldest destroy them which
destroy the earth"*

— (REVELATION 11:15-18, KJV).

Sequence of Events:

1. The seventh (last) trumpet of the tribulation sounds.
2. The earthly reign of Christ begins.
3. The time of the dead (resurrection).
4. Prophets and saints judged for rewards.
5. Those who destroy the earth are destroyed.

Timing: Beginning with the seventh trumpet of the tribulation.

The seventh and final trumpet of the period of the tribulation is sounded and the reign of Christ begins on earth. At this time, the resurrection and judgment of the saints occurs — after the tribulation. This seems to confirm our previous conclusion.

One more verse for further confirmation or refutation:

> *"And I saw thrones, and they sat upon them, and judgment was given unto them: and I saw the souls of them that were beheaded for the witness of Jesus, and for the word of God, and which had not worshiped the beast, neither his image, neither had received his mark upon their foreheads, or in their hands; and they lived and reigned with Christ a thousand years. But the rest of the dead lived not again until the thousand years were finished. This is the first resurrection. Blessed and holy is he that hath part in the first resurrection on such the second death hath no power, but they shall be priests of God and of Christ, and shall reign with him a thousand years".*
>
> — (REVELATION 20:4-6, KJV)

Sequence of Events:

1. The first resurrection (this includes people who are killed during the tribulation for not worshiping the beast or receiving his mark).

2. Those, who are resurrected, reign 1000 years with Christ.
3. The rest of the dead rise after the 1000 is finished.
4. The second death (for those who are not a part of the first resurrection).

Timing: Begins after the people are beheaded for their witness of Jesus, and for not worshiping the beast or receiving his mark, but before the thousand-year reign of Christ.

Here we can clearly see that there will be people who have been executed during the tribulation present at the **first resurrection.** This can only occur if the resurrection occurs after the tribulation. These are the people who reign with Christ for a thousand years. If there is a pre-tribulation rapture, then these people are raised before the first official resurrection, they may not reign with Christ, and they may even be subject to the second death!

Clearly, a pre-tribulation rapture requires that these verses must either be refuted or ignored.

Another scripture frequently used to support a pre-tribulation rapture needs to be reviewed to insure there is no true conflict:

> *"Now we beseech you, brethren, by the coming of our Lord Jesus Christ, and our gathering together unto him, That ye be not soon shaken in mind, or be troubled, neither by spirit, nor by word, nor by letter as from us, as that the day of Christ is at hand. Let no man deceive you by any means: for that day shall not come, except there come a falling away first, and that man of sin be revealed, the son of perdition; Who opposeth and exalteth himself above all that is called God, or that is worshiped; so that he as God sitteth in the temple of God, shewing himself that he is God. Remember ye not, that, when I was yet with you, I told you these things? And now ye know what withholdeth that he might be revealed in his time. For the mystery of iniquity doth already work only he who now letteth will let, until he be taken out of the way. And then shall that Wicked be revealed, whom the Lord shall consume with the spirit of his mouth, and*

shall destroy with the brightness of his coming".

— (II THESSALONIANS 2:1-8, KJV)

Sequence of Events:

1. A spiritual falling away.
2. He who now letteth (hinders or restrains) is taken out of the way.
3. The man of sin is revealed in the temple. The coming of our Lord Jesus Christ and our gathering unto him.

Timing: Just before the middle of the seven-year period of the tribulation when the anti-Christ is revealed.

As I said, this reference is used to show that the church must be raptured before the tribulation. They go directly to verses 7 or 8 and state that he that letteth (restrains) is the Holy Spirit who is in every believer and that when he (the Holy Spirit) is taken out of the way, (so the wicked one can be revealed) those in whom he dwells must also be removed in the rapture. Verse 4 says that the gathering comes after the falling away and the man of sin is revealed. How can this be if the gathering occurs first, removing the Holy Spirit and believers? Obviously it can't.

I found 447 verses dealing with the Holy Spirit but not one reference in which the Holy Spirit is fighting, restraining, or any other adversarial task. The Holy Spirit seems to have three tasks: (1) to comfort believers, (2) to teach believers the meaning of the Word of God and most importantly, (3) convict sinners of their sin. If the Holy Spirit is removed, who will convict the sinners who are saved during the tribulation? If the Holy Spirit is not "he who letteth" then who is?

There are but seven verses about Michael, the Archangel and yet he is fighting against principalities and powers in each one. It is therefore more likely that it is Michael who now restrains the Devil and will do so until told to throw him out of heaven to the earth.

> *"And the woman fled into the wilderness,*
> *where she hath a place prepared of God,*
> *that they should feed her there a thousand*
> *two-hundred and threescore days. And*
> *there was war in heaven: Michael and his*
> *angels fought against the dragon; and the*
> *dragon fought and his angels, And*
> *prevailed not; neither was their place*
> *found any more in heaven. And the great*
> *dragon was cast out, that old serpent,*
> *called the Devil, and Satan, which*
> *deceiveth the whole world: he was cast out*

into the earth, and his angels were cast out with him. And I heard a loud voice saying in heaven, Now is come salvation, and strength, and the kingdom of our God, and the power of his Christ: for the accuser of our brethren is cast down, which accused them before our God day and night. And they overcame him by the blood of the Lamb, and by the word of their testimony; and they loved not their lives unto the death. Therefore rejoice, ye heavens, and ye that dwell in them. Woe to the inhabiters of the earth and of the sea! For the devil is come down unto you, having great wrath, because he knoweth that he hath but a short time"

— (REVELATION 12:6-12, KJV)

Sequence of Events:

1. The woman (Israel) flees into the wilderness (her stay is about 1260 days or about 3.5 years.)
2. Michael and his angels fight a war with Satan and his angels.
3. Satan loses and he and his angels are cast out of heaven down to the earth.

4. Rejoicing in heaven.
5. Sorrow expressed for earth's inhabitants.

Timing: The middle of the seven year period of the tribulation about the time Israel flees into the mountains.

Note that it is Michael who does battle with Satan, so it could also be him whom God has appointed to "let" or withhold Satan's power.

Is there any confirmation that Michael is the one who now restrains Satan's power? Yes!

> "And at that time shall Michael stand up, the
>   great prince which standeth for the chil-
>   dren of thy people: and there shall be a
>   time of trouble, such as never was since
>   there was a nation even to that same time:
>   and at that time thy people shall be deliv-
>   ered, every one that shall be found written
>   in the book"
>
> — (DANIEL 12:1, KJV).

Sequence of Events:

1. Michael "stands up."
2. A time of great trouble (tribulation) begins.
3. Thy people (written in the book) shall be delivered.

Timing: The middle of the seven year period of the tribulation about the time the "great trouble" begins.

So we see here that when Michael "stands up" he delivers those written in the book through a time of great trouble. Since Michael also fights the war with Satan and throws him out of heaven, it is clear that Michael is who restrains Satan now.

Also note that the view of the church being raptured before the anti-Christ is revealed requires you to ignore 2 Thessalonians verses 3 and 4. Let me repeat these verses here for you:

> "Let no man deceive you by any means: for
>   that day shall not come, except there come
>   a falling away first, and that man of sin be
>   revealed, the son of perdition; Who
>   opposeth and exalteth himself above all
>   that is called God, or that is worshiped; so

that he as God sitteth in the temple of God,
shewing himself that he is God"

— (II THESSALONIANS 2:3-4, KJV).

These verses clearly state that the man of sin is revealed sitting in the temple of God before the day of Christ. The reference to II Thessalonians 2:5-6, when taken in context, not only does not support a pre-tribulation rapture, it shows a pre-tribulation rapture is not possible.

Have we covered all the verses used to "prove" a pre-tribulation rapture? No. There are several more references we need to look at. For example Luke 21:36:

> *"And take heed to yourselves, lest at any time*
> *your hearts be overcharged with surfeiting,*
> *and drunkenness, and cares of this life,*
> *and so that day come upon you unawares.*
> *For as a snare shall it come on all them*
> *that dwell on the face of the whole earth.*
> *Watch ye therefore, and pray always, that*
> *ye may be accounted worthy to escape all*
> *these things that shall come to pass, and to*
> *stand before the Son of man"*

— (LUKE 21:34-36, KJV)

I have included verses 34 and 35 to provide the proper context for verse 36. Here a timeline presentation is not helpful. Verse 36 is quoted to indicate the escape of the things that come to pass. The escape is supposed to be the rapture. The problem is that verses 34 and 35 give a completely different picture to this meaning for verse 36. In verse 34 we are told to be careful of ourselves, or we might forget to watch for the coming day because we are too busy surfeiting (eating) and drinking, or that we may become depressed with the day to day worries of this life. We should keep watch or the day will surprise us. Verse 35 tells us that "that day" will come upon all who live upon the earth. ALL!

Verse 36 says to be watchful and pray, so that we will be aware of the events enough so as to avoid the things that will occur, and be able to stand worthy before Jesus. How do we escape? In a rapture, or in like type to the children of Israel when God sent plagues upon Egypt? They went through the tribulations of the day but were unharmed because they were deemed worthy to escape the judgments brought upon Egypt. We are likewise told many times to comfort and strengthen one another with these words. Why would we need to comfort and strengthen unless we will be here?

Another verse quoted is Titus 2:13.

> *"For the grace of God that bringeth salvation*
> *hath appeared to all men, Teaching us*
> *that, denying ungodliness and worldly*
> *lusts, we should live soberly, righteously,*
> *and godly, in this present world; Looking*
> *for that blessed hope, and the glorious*
> *appearing of the great God and our Savior*
> *Jesus Christ; Who gave himself for us, that*
> *he might redeem us from all iniquity, and*
> *purify unto himself a peculiar people,*
> *zealous of good works"*
>
> — (TITUS 2:11-14, KJV).

Again, I add the verses around the referenced one to look at the context. This verse does indeed speak of Jesus return, but it doesn't provide any idea of when to expect this event. It simply tells us we should always be looking with joyful anticipation so that we no longer are tempted by ungodliness and worldly lusts.

Are there any other verses used to support the pre-tribulation rapture that need to be considered? Yes, we need to deal with the idea that Christ will come upon us "as a thief in the night." This comes from I Thessalonians 5:2. Again I will include additional verses to ensure this verse is not taken out of context.

"But of the times and seasons, brethren, ye have no need that I write unto you. For yourselves know perfectly that the day of the Lord so cometh as a thief in the night. For when they shall say, Peace and safety; then sudden destruction cometh upon them, as travail upon a woman with child; and they shall not escape. But ye, brethren, are not in darkness, that that day should overtake you as a thief. Ye are all the children of light, and the children of the day: we are not of the night, nor of darkness. Therefore let us not sleep, as do others; but let us watch and be sober. For they that sleep sleep in the night; and they that be drunken are drunken in the night. But let us, who are of the day, be sober, putting on the breastplate of faith and love; and for an helmet, the hope of salvation. For God hath not appointed us to wrath, but to obtain salvation by our Lord Jesus Christ, Who died for us, that, whether we wake or sleep, we should live together with him. Wherefore comfort yourselves together, and edify one another, even as also ye do" (1Thessalonians 5:1-11, KJV). Yes, Paul writes that the Thessalonians themselves know that the Lord's return will be a thief in the night. But verse 3 indicates that believers are not "children of the night" but are "children of the day" and should not be surprised at the Lord's coming. Is this meaning one we can confirm?

Yes, we can see the same theme expanded on in Revelations 16:15.

> *"Behold, I come as a thief. Blessed is he that watcheth, and keepeth his garments, lest he walk naked, and they see his shame"*
>
> — (REVELATIONS 16:15, KJV).

What does this verse mean? A Jew would recognize the reference to the thief in the night catching one sleeping and losing their garments. Let me explain. Each night in the temple, one priest was left to tend to the altar to make sure it did not go out during the night. Each morning the High Priest would return just before dawn. It was said that he would enter the temple as a "thief in the night." His first duty was to check the altar fire to see if it had gone out. If it had, the High Priest would relight the altar fire (only he could) and would take some coals in a fire pan and go looking for the sleeping priest.

When the High Priest found him, he was to awaken him by lighting his clothing with the coals. The priest would awaken with his clothes on fire and they would burn very fast. It was therefore said that he would have to walk to his home naked and that everyone could see his shame. This certainly does not sound like we are going to please our Lord if we are surprised by His return.

By now, you are probably saying: so why tell me this? Why do you take up a discussion that will probably only be answered when the event occurs? Let me explain my reason for writing this booklet.

Many good Christian pastors, evangelists, and teachers teach the pre-tribulation rapture as fact, not just one of several opinions. Many have been taught it as fact themselves and believe it to be so. I myself was also taught it as fact and for many years believed it.

Many logical reasons and theories, based on a number of assumptions were provided to support the pre-tribulation rapture. None were totally based on scripture and so, I was never quite sure. I said that as long as I could find no scriptures that told me for sure when it was, that I would believe in a pre-tribulation rapture.

Some time ago I read some scriptures nobody had ever quoted regarding the rapture and they were clear and they were specific about when this event was to be. I've quoted some of those verses here.

What really concerns me is that some pastors and teachers may not believe in a pre-tribulation rapture, yet feel compelled to teach it anyway. This may be so because some of them may believe that if they teach their church members they will go through the tribulation, many members will be offended, possibly placing his job in jeopardy.

But I believe that the doctrine of pre-tribulation rapture is the strong delusion to be sent according to II Thessalonians 2:11. If you believe in the pre-tribulation rapture, you are not preparing to withstand the tribulation challenges. You're going on a trip; A trip away from all the troubles of the time! The Jews must face this time alone. Anyone else can escape by being saved before the tribulation begins. Since the only sign that occurs at the beginning of the tribulation is a peace treaty in Israel with a seven year duration, the first time you will have any real evidence that the tribulation has begun is when the Abomination of Desolation stands in the Holy Place.

If you are here on that day, will you have kept yourself from the snares of the Devil? Or will you also begin to wonder the question:

> *"Knowing this first, that there shall come in the last days scoffers, walking after their own lusts, And saying, Where is the promise of his coming? For since the fathers fell asleep, all things continue as*

*they were from the beginning of the
creation"*

— (II PETER 3:3-4)

Look at this text. Some say this refers to scoffers and weak believers who have been waiting for the pre-tribulation rapture when they realize the Abomination of Desolation has been revealed. They will turn to their teachers and perhaps even their pastors and ask: Where is the promised pre-tribulation rapture or literally the coming of Christ for His church?

Won't you please consider these quotes and search the scriptures yourself? Read the verses and check the context of the verses. Ask God to show you what is true concerning this subject. There are scriptures that tell us to comfort and strengthen each other with these words, just as the children of Israel were told to be strengthened and see the hand of the Lord at work in Egypt. We too are to remind each other that God will deliver us from the plagues coming upon the earth. Will you be ready? Will you have made proper plans?

- End of Jeff Mott's Article -

Always the Bible, Only the Bible, and Nothing but the Bible.

Christ will appear in all his glory with his angels and *"will gather his elect from the four winds, from one end of heaven to the other"*(Mat. 24:31). Paul details this gathering using graphic resurrection and rapture language (1 Thess 4:13–18). Between the sixth and seventh seal, the book of Revelation depicts the raptured as *"a great multitude that no one could number"* these will be *"the ones who have come out of the great tribulation"* (Rev 7:9-end). This event will be **immediately followed** by the start of God's judgment (Rev 8:1–6). We are not appointed to God's wrath. Jesus, Paul, and Peter warn these judgments are for the ungodly (Matt 24:37–44; 1 Thess 5:1–11; 2 Pet 3:1–18).

It is important for us to continuously and humbly scrutinize the Bible, to understand what the purposes of the Lord are. The doctrine commonly called the rapture, as defined in I Thessalonians 4:16-18, is given by God to provide hope to those steadfastly enduring the great tribulation. Again, it is written: *"For the Lord himself shall descend from heaven with a shout, with the voice of the archangel, and with the trump of God: and the dead in Christ shall rise first: Then we which are alive and remain shall be caught up together with them in the clouds, to meet the Lord in the air: and so shall we ever be with the Lord. Wherefore comfort one another with these words."*

A worthwhile goal for us therefore, is to endure and to stay *"alive and remain"* through whatever occurs, until we finally *"shall be caught up together with them in the clouds."* Let us *"comfort one another with these words."* Be encouraged, for though the wrath of satan Rev. 12:12, will have extreme and testing global events, Christians know their eternal rewards are absolutely secure.

"Blessed are the dead which die in the Lord from henceforth: Yea, saith the Spirit, that they may rest from their labors; and their works do follow them" (Revelation 14:13).

For us it is certain, whether we 'sleep' or *"are alive and remain"* through to the end, we will finally *"meet the Lord in the air: and so shall we ever be with the Lord."* Believers are *"persuaded, that neither death, nor life, nor angels, nor principalities, nor powers, nor things present, nor things to come, nor height, nor depth, nor any other creature, shall be able to separate us from the love of God, which is in Christ Jesus our Lord"* (Romans 8:38-39).

God's love and salvation for us are absolutely permanent, never obstructed, and surely abundant now and in our future!

Before the *"snare"* is upon us, people/leaders who have publicly promoted the pre-tribulation rapture idea, would be wise to **share that this is inaccurate, and that not everyone accepts the assumptions as settled doctrine.** Then use whatever platform they have to clarify God's written direction. The Bible describes a tragic *"falling away"* that will occur, before the man of sin is revealed. So, Christians will recognize the end has begun **somewhat earlier,** see the order in II Thessalonians 2:1-8. The great tribulation is not a time for Christian leaders to sidestep biblical events, or present spurious doctrines to protect their time-honored positions, as did those hardened ministers of Jesus day. The people over whom a deceived pastor has oversight, will be very upset at having been misled. Make no mistake, God will not lightly judge leaders if they selfishly dig in their heels and misguide or promote bad doctrine at this difficult time.

Having believed and promoted the pre-tribulation idea myself for years, I ask you, **please forgive me**. It seems unavoidable nevertheless, that such deeply disappointing news will tempt many to scoff at **Christians in leadership, and even be angry with God Himself**. Of course this should never be. Remember, only the Lord could have allowed such widespread acceptance of a false view among His people.

Joseph, in a position to get revenge on his brothers who callously sold him as a slave, instead lovingly forgave them. He mercifully claimed, *"God did send me before you to preserve life"* (Genesis 45:5). Now please recognize, no one has been callous or deceitful in promoting the pre-tribulation belief. Forgive quickly every leader in your life who misunderstood the sequence of the Lord's coming. Just as God allowed Joseph's brothers' hateful actions to position him where he would finally save them from a famine, let us view this second coming sequence confusion as a similar permissive act of God's mercy...

Just think how uncomfortable every generation might have lived over the decades, subject under a looming great tribulation shadow. Or, as with human nature, we might have gotten relaxed with the knowledge, and then done little to prepare. Or possibly, even more destructive, we might have centered only on preparing, only to isolate ourselves, all the while living selfishly useless for God's ministry purposes. And ultimately end up looking nonsensically fear-filled to the unsaved to whom we were supposed to be sharing the Gospel.

Enough of speculation. We rest in our Lord who has his infinitely higher reasons; who also knows how to preserve and timely transition our focus when required. This is His earth; we are His people in submission under His leadership - a comforting fact, whatever our future holds.

Putting all hindsight confusion, angry accusations and offenses aside, we know that God is sovereign in all matters. May I be blunt... **pass your test; quickly forgive anyone who might have confused you. Turn and move forward to meet the challenges we now face.** Bottom line, as we look forward, *let us do better* facing our trials and testing than our historical forefathers did with theirs. It would be tragic if the centuries of ever-increasing abundance that our heavenly Father has heaped on Christians around the world, finally resulted in a bumper crop of spoiled, self-centered adult brats, who refuse to continue humbly following and worshiping Him, because the blessing spigot has slowed to a third-world trickle. Many people in foreign countries are grateful when they lay their heads down at night having simply eaten a meal with clean water. The Bible says, with food and clothing we are to be content. 1Tim. 6:8

How shall we cope going from Little League practice one month, to hauling water from a nearby lake a few months later, to avoid the mark of the beast, required to pay the water bill? Make no mistake, the devout Christian will joyfully continue to do justly, love mercy, and walk humbly, with his eyes on the prize, to enter the presence of our loving Lord Jesus and vast multitudes of trouble-free saints, in joyous eternal glory! Rev. 7:9-17

Many people have debated the meanings of the biblical "great tribulation" with its unique metaphors and unclear sequence of events. While our Lord always knows what we do not understand, He can as well, unfold revelation as He chooses, at times and in ways that fit His purposes. I pray God will use this book somehow in that regard.

Keeping ourselves down-to-earth with a plain biblical context, it will help our understanding if the seven year era is overlaid with the prophetic four horses of Revelation, which appear as the first four seal judgments (see Revelation 6:1-8).

This era begins with a quiet rise to power of a white horse ridden by a newly crowned conquering ruler - during the first three and a half years, as the first seal is opened (vs. 2). Then, a global financial collapse will occur that is characterized solely by tremendous bloodshed worldwide, as the red horse sweeps over the earth, when the second seal is opened (vs. 4).

Finally, the last horses of the apocalypse appear: the black one races over the planet with the opening of the third seal (vss. 5-6). This is a metaphor for starvation across earth. Then the pale one rides, as the fourth seal is opened, indicating a time that death will occur to one-fourth of the people.

Somewhere within these four horses or calamitous global phases, the Lord confirms for His people a singular highlighted pivot-point in His overall timing. He mercifully distinguishes for us when the start of the most chaotic time begins, about midway, calling it the "great tribulation" (Matthew 24:21). Here begins the Daniel chapter 12, 1335-day (about three and a half year) countdown to the final end of the world.

Beginning the count and indicating day one, is an abomination of desolation event.

We can be encouraged, for God tells us the duration of this coming traumatic season. **The countdown begins from the day the daily sacrifice is taken away and an abomination event occurs in Jerusalem, according to Daniel 12:10-12**:

> *"Many shall be purified, and made white, and tried; but the wicked shall do wickedly: and none of the wicked shall understand; but the wise shall understand. And from the time that the daily sacrifice shall be taken away, and the abomination that maketh desolate set up, there shall be a thousand two hundred and ninety days. Blessed is he that waiteth, and cometh to*

the thousand three hundred and five and
thirty days."

— DANIEL 12:10-12

Our Lord is showing to us His abundant mercy, as He specifies for Christians worldwide, the great tribulation starting day and the complete duration to be 1335 days. Knowing about this period will allow prepared people to meter out food and their other limited stored provisions to endure. What wisdom of God; we are never in the dark! When we see the sacrifice stopped, that is when we can identify the anti-Christ with absolute certainty. More importantly, we are told to start counting days from **that day**, according to the above biblical instruction. Before *"the man of sin be revealed,"* however, as II Thessalonians 2:1-3 warns, a *"falling away"* will occur among believing people who once walked in God's way.

Dear friend, hell is certainly too hot, eternity is way too long, and heaven too wonderfully glorious, to consider *even for an instant*, damning yourself, by taking the mark of the beast, to eat. *"Then a third angel followed them, saying with a loud voice, "If anyone worships the beast and his image, and receives his mark on his forehead or on his hand, he himself shall also drink of the wine of the wrath of God, which is poured out full strength into the cup of His* *indignation. He shall be tormented with fire and brimstone in the presence of the holy angels and in the presence of the Lamb. And the smoke of their torment ascends forever and ever; and they have no rest day or night, who worship the beast and his image, and <u>whoever receives the mark of his name</u>."* (Revelation 14:9-11)

Dear Christian, Do NOT take the mark of the beast! DO NOT FALL AWAY TO BRIEFLY EXTEND YOUR OWN LIFE; EVEN IF THOSE YOU LOVE FALL AWAY! No spouse, or Pastor, children, friend, family, group, or doctrine, is worth you personally suffering for ALL eternity. I wish I could look you in the eyes, and say: ***Make your own decision – make it unshakable, personal and permanent. Vocalize it to everyone; even promote the decision far and wide... proclaim that you will not take the mark, even if martyrdom is required! Now, you're being both heroic and courageous!***

*Decide that you will NOT take the mark of the beast– no matter what, speak of your reasons to **everyone**, and stick to that decision, as if your eternal life depends on it- regardless of how **any** others respond or how your earthly life may become disgraced, degraded, despised or destroyed.* Wise people, if they learn of the need for food early enough, will make plans and preparations to endure. See Mt. 24:44-46. This stressful brief time is painted in the Bible as nearly unbearable. Remember, your central and exclusive treasure now is not your bank account, your job, house, or anything on this declining crumbling earth. Your salvation was purchased by the precious blood of Jesus on the cross -it is for you! This salvation is your and my only remaining wealth.

OH, THIS TREASURE WE HAVE IN HIM!!! LOCK YOUR HEART ON HIS SALVATION PROMISE, AND THE JOY THAT IS SET BEFORE YOU... AND *NEVER LET GO*! Remember this challenging thought, per the Bible, every person on earth will die in the last three-and-a-half year, great tribulation period. Your and my goal (but not at the expense of hurting innocent people) will be to endure, or if need be, to enter God's 'sleep'. We must stay focused on retaining the purity of our hearts in Christ, and abounding in the fruits of the Holy Spirit, through to the end. *"and because iniquity shall abound, the love of many shall wax cold. But he that shall endure unto the end, the same shall be saved." Mt.24:12-13*

Read Hebrews chapter 11. Every challenge to a believer's faith can be met successfully, particularly when it ends with early 'sleep'. We rest confident; God's grace will *surprise* and *strengthen* you and me to make good choices. Of course, He knew you and I would live during this time. God is looking forward to our loving service to others, and to our calming countenances, to support and encourage our loved ones!

The Holy Spirit is in us today just as He was in men and women of faith in Hebrews 11. We, likewise, look to God's Word to provide plenty of courage and guidance for us. With our earthly finish line just on the horizon, we must not lose heart or quit life's race. Many worldly living in the flesh, will surely escape to heavy drinking, engage in promiscuity, do drugs, etc., for their comfort. The ignorant and those who do not seek truth will turn to their wickedness. Steadfastly endure in the faith; we do **absolutely win**! God has let us know a full 2000 years ago, about all of this. Now it's showing up in front of us.

Have you ever wondered if God occasionally rolls his eyes at what he sees people doing on earth? If our panicked response causes fear-to-paralysis, when we see the ancient prophesies go live, *this* - it would seem, would make him want to. Surely, we can take heart that we have been chosen by God to live at this time. We are not accidentally here now. Our Lord has confidence that you and I will look to His Word, seek his face, and continue to walk in the fruits of the Spirit.

Think about it, the very fact the great tribulation itself is actually happening (I write this by faith; at this moment- Nov.5, 2022, I see only a false president and general decay of Christian values). We should view the great tribulation when it starts, as monumental biblical confirmation, that God loves all people enough to allow - at the closure of the world, a mandated eternal life-or-death choice.

> *"For whosoever shall call upon the name of the Lord shall be saved!"*
>
> **— ROMANS 10:13**

Everyone is mercifully compelled with a chance to choose salvation! Understand, Salvation in Jesus is all that matters... afterwards begins everyone's eternity, and that will never change!

> *"It is appointed unto men once to die, but after this the judgment"*
>
> — HEBREWS 9:27

The next chapter will reveal some remarkable things that are happening now and open a door into a warehouse of biblical evidence for God's final human closure.

3

WHAT IN THE WORLD IS GOING ON NOW?

With the judgment and anxiety of our sin removed from us, by the finished sacrifice of Jesus-himself, on the cross, we are not slated for God's wrath in the future. The Lord lovingly watches each Christian with only desire to reward even our simple deed of giving a cup of cold water. Much more our right responses to eternity-impacting doctrine. Yet, we each must transition to heaven through death, or alive in the final catching-away. Please do not be reactive but stay prayerful as you read the remainder of this book.

The Lord Said He Will Change His Handling of Mankind

The Lord desires that all people would be saved. He has delighted to satisfy the souls of everyone who has turned to Him, all through the ages. Almighty God has, with great long-suffering, endured centuries of pride, lust, and selfishness of a headstrong human race. He has always loved wayward people having died for us while we were still sinners.

To make possible many billions of additional 'sinning' people in the future, God scattered the population out worldwide, by changing languages, after the Flood. This prompted people to separate around the world and multiply, just as we see today. The Lord continues to yearn for each individual, and every generation, to recognize their disastrous state of sin, learn of His forgiveness available, and embrace like a child, His simple gift of salvation. The Lord's decisive wish is that people the world over begin a joyful personal relationship with Him! As we learn the value of our salvation gift, and experience the joy of walking with God, old desires to sin become repulsive. Thus begins our upward life of sanctifying loyalty to the Lord.

Every person has varying ability and desire to learn new information. Again, please be peaceful in your responses to the remaining material in this book. Structure words of faith as you study these things. I am a man with a family; these insights have been tough for me to contemplate as they may be for you. Stay in a faith-filled trusting frame of mind toward our loving Father. *"let the weak say, I am strong"* Joel 3:10

Our Redeemer is forever loving and gentle to His own, even if our surroundings turn chaotic. Remember that we, believers, are the apple of His eye. Our names are carved in the palm of His hand, according to Scripture. Since creation, our loving Lord has provided a narrow but sure pathway for any pilgrim to step onto and receive salvation, along with His joy and peace. Bottom line, God wants you and me in heaven with him!

Christians of old did not always see the fulfillment of their dreams completed during their earthly lives. Should the great tribulation disrupt our lives, we may not see all our earthly desires fulfilled either. People with hearts rooted deeply in a rapidly collapsing world, like Lot's wife, will undoubtedly be vulnerable by their wrong passions, as she was. All steadfast believers however, will most certainly realize the Lord's ultimate plans, fully, and in unimaginable ways in heaven. It is written: *"Eye hath not seen, nor ear heard, neither have entered into the heart of man, the things which God hath prepared for them that love him"* (I Corinthians 2:9).

Scripture declares that at God's right hand, there are pleasures for evermore. First, however, we are given to understand a purifying decision-demanding transition must take place on earth. "*As a snare*," there will be an abrupt end to this current peaceful age of relaxed grace. Your Bible describes a brief 1290-day change in God's primary handling of all men living, to that of closing judgment. This short era will be followed by an end-all wrath upon remaining hardened God rejecters, according to the books of Daniel and Revelation; more on that later.

Briefly looking back in history, we see that having accurate guidance alone was not enough for people to succeed with the Lord. A call from God unheeded led to the drowning of all but eight people of Preacher Noah's day. Even more tragic, Moses' followers had clear guidance and saw daily miracles, yet only two men and their families, of the original jubilant set-free multitude, finally entered the Promised Land. These are examples to us, of how with a scoffing attitude or fearful outlook, people can reject God and His favorable pathway, resulting in an untimely loss of life.

Almighty God emphasized strict obedience of His Law(s) in the Old Testament; then in the New Testament, since Jesus came, we have been led with the gentle rule of His grace and His mercy. Thus, God has provided for all believers freedom to accomplish an important harvest, to go worldwide and make disciples (Mathew 28:19). All people entering the tribulation will experience another brief worldwide focus, that of a trying choice. *That is, to value and retain our eternal life.* God will intensify His dealing with all individuals, by allowing a forced decision for or against Himself. It will be a pivotal time. Those comfortably active in sin will naturally react from their flesh, without an ear for God's guidance. Truth-seekers will be asking God how to respond, while searching His Word, and staying sensitive to the Holy Spirit.

In my desire to encourage you, if it were practical, I would write the following verse across your backyard, stacking concrete cinder block letters three-foot high:

> *"Looking unto Jesus the author and finisher of our faith; who for the joy that was set before him endured the cross, despising the shame, and is set down at the right hand of the throne of God"*
>
> — (HEB. 12:2).

It was His personal anticipation of heaven's eternal joy, chiefly, that sustained Jesus in the torturous events He went through. *That same joy is awaiting you and me!* We must never undervalue the worth of the eternal joy that awaits us. It will motivate us on to victory. Think about it, meditate on it, and never forget it. *You have indescribable joy awaiting you in heaven, and it will last for all eternity! God wants this for you and me!*

God Wants Us to Know, What He Wants Us to Know, When He Wants Us to Know It, and Only Then May We Know It!

The Lord's use of prophecy has worked in the past, as an alerting and confirming tool. His signals were prepared and aimed for future people to recognize, accept, and wisely respond to approaching world transition. In an early prophecy, the Lord told Satan: Eve's seed *"shall bruise thy head, and thou shalt bruise his heel"* (Genesis 3:15). We see that God had prepared this obscure sign way before time, intending that the applied meaning be held off thousands of years, until it plainly fit some event in the future. In this case, Jesus' crucified heal dealt an effectual death blow to Satan's head.

The verse serves as a key to unlock for those onlookers present, a new understanding. It introduced them to a major surprising era - the entry of the Age of Grace (unmerited favor), and a whole new method of living. For them and now us, it means that we can live redeemed, with our sin debt paid. We embrace eternal life through simple faith in the single death and resurrection of Jesus, no more animal sacrifices. What's more, we live now without any condemnation of sin: past, present, or future (Romans 8:1-2). We also experience the joy of freely sharing this new life with others!

In contrast, right at this moment, a different key-to-an-era verse appears to be activated. The book of Revelation holds a very different pivotal passage for the world to recognize change coming. It directs attention, not to the sacrificing Christ, but for all people to identify a coming counterfeit leader. The Lord intends for people everywhere to identify an anti-Christ candidate. God has targeted this recognition to alert early the approaching arrival of a brief intense, decision-demanding time. Consider this seemingly random command given by God in Revelation 13:18, *"Here is wisdom. Let him that hath understanding count the number of the beast: for it is the number of a man; and his number is six hundred threescore and six"* or 666.

As people faithfully obey the Lord and personally count, it brings to mind a realization of a completely new era of human time and direction on earth. As we dig further into the implications, a whole panorama of supportive evidence comes into view, revealing the probable nearness of the great tribulation judgment. By obeying Revelation 13:18 to personally count, we will see the Lord unlock a new door into a bounty of Scriptures which match with the prophesied final closure of man's time on earth. When this most challenging period begins, we are to cling to our Lord by faith, and endure with our faith intact, at any cost. The Church, must of course, continue to provide the life-changing Gospel as we always have.

In addition, the Bible instructs that we are to unfailingly, *"Let"* others count *"the number of a man."* This command to let others, is prefaced with a motivating label: *"Here is wisdom..."* Imagine "Here" as a doorway, and each believer who counts – enters, and becomes knowledgeably positioned to expect and is confirmed with fulfilled prophecy in their mind, to respond to the foretold arrival of new waves of world change that our Father has said will follow.

Again, all Christians living at that time will enter the great tribulation. How sad for those who enter with the idea, '*because I have been "Left Behind" I am counted among the lost - I missed the rapture!... the Holy Spirit has left the earth...*' You must know this is a false doctrine. That horror could lead people to actually fall away, in despondency. Any anger, bitterness, blame or disgust toward Christian teachers and leaders will only add to their confusion. The pre-tribulation rapture doctrine is

100% assumption-based. It is **not supported** with direct verses anywhere in the bible. It was wrong when it was popularized 190 years ago by John Nelson Darby, and more recently by Tim Lahaye and Jerry Jenkins. It is still wrong today.

If you find yourself in the great tribulation, and you have previously accepted Jesus, **be encouraged**, the following verses are given by God to comfort us. Highlights indicate the **great tribulation context**: *"But I would not have you to be ignorant, brethren, concerning them which are asleep, that ye sorrow not, even as others which have no hope. For if we believe that Jesus died and rose again, even so them also which sleep in Jesus will God bring with him.* **For this we say unto you by the word of the Lord, that <u>we which are alive and remain unto the coming of the Lord</u> shall not prevent (go before) them which are asleep. For the Lord himself shall descend from heaven with a shout, with the voice of the archangel, and with the trump of God: and the dead in Christ shall rise first <u>Then we which are</u> alive and remain shall be caught up together with them in the clouds, to meet the Lord in the air: and so shall we ever be with the Lord. Wherefore comfort one another with these words."** *1 Thess. 4:15-18* **Be comforted!**

Certainly God's Holy Word will continue to direct us as we seek Him to make wise personal choices. The Lord's model for past transitioning people, as we discussed, was that they match their surroundings to His Word and so retain **their all-important orientation within Scripture.** This focus is for us as well. Doing so, our priceless faith will be increased, decisions are better navigated, and we can more willingly trust God with any and all outcomes, as we pass through our assigned transitions. We, of course, identify our Savior's first coming, looking back at His completed earthly life. We accept His death and glorious resurrection, as written for us in the New Testament, and as they relate to the many prior written prophecies.

Jesus' Second-coming indicators, symbolic *"horses,"* *"bowls," and "trumpets,"* can be puzzling, but were never intended to be conveyed via movie fantasy or whimsical book fiction (accessible only to a very small portion of earth's people). God does not act like a mystic, nor is He flippant with a hide-and-seek delivery of knowledge for fun, or to taunt or show off His superiority. Symbolic descriptions and metaphors are the Lord's brilliant way to withhold, preserve, and thoughtfully deliver closed bundles of knowledge forward, to his targeted moment in which relevant activity, people and places, plainly match Bible signs - for the good of those who will need the official reference and certified understanding with confirmed recognition.

Many of these God-given symbols are aimed to expose the anti-Christ. This makes unwrapping this subject of study somewhat uncomfortable to a believer's sensibilities, rather like gossip. However, as Bible-embracing Christians, we are committed to the unflinching study and acknowledgment of the **whole** of what the Scriptures say to us, even if reading, we feel uneasy.

Cherry-picking doctrine solely as it provides emotional comfort, is a dangerous and tragically common practice. As things begin to rapidly change, selective tunnel vision, allegiance to a conflicting denomination, or a self-serving prominent/charismatic leader, could even pit family members against one another. Don't allow anyone, any doctrine or any public expectation to steer you against the Bible, into confusion, or into eternal destruction.

What if some current activities today were to apply to us and show early indications that the great tribulation is on our horizon? Many particular elements of our surroundings would begin to match up in obvious and unmistakable ways with God's biblical road signs. The next chapter will turn to focus on this possibility. Let us use the Bible now to define the Bible, and see if it applies to what is going on around us today.

The wise men of old were diligent to pack up and travel as they did, to see Jesus at his first coming. Contrasting their zeal to Christians today, are we as eager to respond to biblical signs of our Savior's very different second coming event? In our Lord's last moments, Jesus pondered this out loud. He vocalized if when he returned he would find faith in **anyone**: *"when the Son of man cometh, shall he find faith on the earth?"* Lk. 18:8. The question is not frivolous or meant to provoke a flippant response: "Of course, I'll have faith!" No, instead, we should humbly ask, why does Jesus sound so doubtful; why would he ask such a bleak question - whether faith would exist anywhere in anyone on earth when he returns?

What in the bible hints of a reason faith will become so difficult for Christians, that Jesus would contemplate and highlight in scripture, such a bleak worldwide setting? First, scripture forewarns of a time when massive convincing lies will sweep multitudes of people away from God. Today, off-the-shelf AI, artificial intelligence software, can be bought and used to generate convincingly fake videos. For example, a president or any person, can be digitally made to say anything you want them to. Coupled now with a bought-out mainstream media, wholly controlled and comfortable with cultivating false narratives – even unlimited crafted lies, our world is equipped for seamless and realistic audio-visual global deception.

Anything that corrupt minds might imagine - anything, can be made to look absolutely real with computers today. With AI software, whole cities can be made to look on fire with virtual people panicking. Flying saucers could be made to land with aliens exiting, walking and talking. Jesus can be made to appear, with realistic glory - AI generated and 100% fake. Any idea can be made to appear real with literally no limits. Follow this up with the primary news agencies backing the stories with graphic details, as they pull aside to 'interview' weeping actors with testimonials - and *most people in the public* will believe whatever they see and hear. Look for the lie, that Jesus has returned to earth: *"you must go see him!"* Of course, we must not be swayed.

Standing ready right now, the global media is united, poised and fully equipped to present any message, as persuasive polished news. (web search: "Elon Musk, AI danger, releasing the demons.") Christians must cling to the *biblical Jesus* with peaceful trust, and disregard anything *seen or heard that persuades contrary to the plain writings of the bible.* Will Jesus find faith in you and me, when he returns? It is a real question. Establish in the core of your being, that you will not take the mark of the beast, no matter what you hear or see, from this moment forward. We pray, God will give discernment.

The bible warns as well of a very real fundamental global change on the horizon. Physical money will become useless. President Biden recently signed Executive Order 14067. The clear intention is for banking to become all digital and controlled by the government. Soon, every American bank account could be subject to funds being frozen, removed, managed, and 100% monitored. Government managers with a key stroke and algorithms, may instantly remove or limit anyone's financial independence based on their purchases, giving, political ideology, or for any reason at all. Virtual cash may begin with a national catastrophe – the most effective method for the cabal to focus blame. Even right *now* we are breath-holding to see if the national debt ceiling will be timely negotiated. Dangerous brinkmanship is playing out in DC that could function for them *"as a snare"* to collapse our economy, and the world's. Around June/July of 2023, in an unknown moment, money to pay government debt is set to run out. Confidence in the US dollar - the default global currency, would be shattered. This, or a similar event may be the start of the Revelation Red Horse era.

"As a snare shall it come on all".

— LK. 21:35

Believers should keep a healthy suspicion of most news presentations today. For example, were three March, 2023 bank failures clumsy coincidental mishaps, or all in the same week, strategic? Treasury Secretary Yellen quickly trotted out the soothing announcement, that protective FDIC bank limits are removed. Suddenly, no matter how large the amount of money in the foundering banks accounts, no one had any risk. Was this done to groom and pacify the financial sector to prevent bank runs prior to the Biblical "snare" dollar collapse? It is wise now to watch all news with doubt and questions. Be alert to the *effect* of what you see and hear, on yourself and others.

Each person who takes the mark might become debt-free and given virtual piles of digital cash, perhaps based on their level of commitment upon entering. An extra bonus of "digi-cash," could be gifted to those who refer their neighbor... (or a triple bonus awarded for a confirmed report that a neighbor is avoiding the system? Betrayal and treachery are biblical warnings). Good marketing usually comes with a cheerful byline, maybe something like, "Evolve with the Mark!" These ideas are imaginary, the Executive order is not. The Bible is literal and factual and completely accurate. The mark will be offered - possibly required; however we will *never participate; God said not to, and we CERTAINLY WILL NOT!*

If your family hasn't eaten in weeks would you be tempted to take the mark? Pause and read Revelation 14:9-13, out loud. It warns of eternal fiery torment for *everyone* who takes the mark of the beast. Talk about stress... how about another of your family members - to stop the suffering hunger of children? Pressure against believers could tragically turn some to take this damning mark - even knowing the result will end with their fiery eternal outcome. Sadly, their decision will be for their own destruction forever... NOT SO FOR YOU! You have the Holy Spirit living inside of you. The mind of Christ will certainly guide Christians; we adhere to His Word, treating every passage as weighty as a salvation doctrine. We live not just by food, but remember, by *"every word."* The Bible is ALL God's Holy Word.

In the end, each person's biblical adherence, or their individual lack, will echo throughout eternity. Like Jesus, we have unspeakable joy set before us! Our Lord gained strength looking forward to heavenly joy in front of him. **So must our coming heavenly joy, also become our personal steadfast motivation and our single highest valued objective on earth, bar none.**

Repeatedly, the bible declares, God's judgment is coming upon the whole world, but why would this be? Let us consider (though surely not actually possible) from God's viewpoint, one single horrific fact of our day… millions of gallons of precious innocent babies' blood has been violently shed into the world's sewer systems, over decades. The many tiny torn lives were intended by their Holy Creator, to live, to love, to learn, to create, to enjoy, to invent, to play, to smile and laugh, to cuddle, to walk and talk and discover God's beautiful world. They were supposed to leave their joyful unique influence and mark, for others to benefit. God deeply wanted them each to live as you and I have been so blessed with life. They were each likewise intended to reach a good old age. Evil laws have allowed legal massacre, all over this planet.

Almighty God has always punished murder, starting with Cane. To date, just punishment for this vast sea of stolen lives has been withheld. This culturally accepted child carnage, links us to a pivotal signpost of prophecy. The verse: "as it was in the days of Noah" Mt. 24:37. What was in the days of Noah? Genesis 6:11, *"The earth also was corrupt before God, and the earth was filled with violence."* I hope you are seated for these next sentences…

According to www.usabortionclock.org over 1,600,000,000 children have been aborted globally, since 1980. For perspective, if each missing person were given two linear feet to stand in, the unbroken line of murdered people made in God's image, by abortion, would stretch more than 606,060 miles...

This line of very real people, would extend unbroken around the earth twenty-four times! These are individual persons who have been robbed of their life. Friend, if just now, reading that statistic you did not weep or feel deep emotion, ask God to open your heart to the wicked reality of 'our Noah-like day'. Contrary to the claims by the media, the abortion count required to save a life, is very small; statistically a percentage of zero. I will list details only of abortion for emphasis in this book. You well know our world is ravaged by sin of every kind, sex-trafficked children, drugs, corruption in governments, the list and statistics would go on for pages.

Astonishingly however, God has met our decades of violence, perversion, greed and pride, with prolonged silence. He has postponed retribution, and has ultra-patiently withheld delivery of our century's-deserved judgment. Maybe God is enabling a final calm, for a last saving gospel call? Not much else can be imagined. Judgment is uncomfortable to think about and not popular, especially from the pulpit today. Modern American Christian culture has difficulty discussing God as our judge. More directly, it's not a pew-filling topic. Over time, it is natural that believers have become numb to these gruesome abortion and broader sin facts. The carnage was done quietly, discreetly, legally, using sterilized "health care" equipment, with soothing music in the background.

An argument can be made, that American Christianity is unfocused about judgment, even while all mankind blindly teeters on the precipice of Revelation's end-of-world descriptions. Have you ever heard a sermon from Revelation, about: *"in (those) days shall men seek death, and shall not find it; and shall desire to die, and death shall flee from them"?* Rev. 9:6.

We focus almost exclusively today on God's wonderful Grace, and truly it is glorious. We repeatedly stress his Mercy - and how grateful we each are for Jesus' sacrifice! Preachers assist us with how to fulfill our dreams. These are encouraging biblical subjects. Neglected today though, is candid preaching, or even open discussion of major end-of-the-world issues (topics the disciples and Jesus casually discussed). Excluded in modern Christian dialogue are the various judgment verses with timing and context questions, related to the end of the world. But frankly, how is a pastor to deliver a message on Rev. 6:8: *"And power was given unto them over the fourth part of the earth, to kill with sword, and with hunger, and with death, and with the beasts of the earth."* A fourth part of the earth today is 1,975,000,000 people, dead! Where do we file that sermon walking out of church? It is information, in the Bible, given for believers to be aware, that an extremely difficult season characterizes mankind's closeout time on earth.

Or, how about what is perhaps the most unpopular of all bible messages, that Satan will briefly overcome everyone in the end, including all Christians worldwide: *"And it was given* **(by God)** *unto him to make war with the saints, and to overcome them: and power was given him over all kindreds, and tongues, and nations."* Rev. 13:7 These need not be part of our everyday meditations, though they are certainly relevant biblical truths for an appointed time in the future. They will be impacting tests for those living at that time. These are verses that we should not completely ignore, as has been done.

Jesus declared, we will live *"by EVERY word that proceedeth out of the mouth of God."* Mt. 4:4 If the above intense verses are permanently neglected, doubtlessly the result will be undue shock among believers when the great tribulation begins, particularly since the majority are expecting a non-existent pre-trib rapture. A rapture concept, whereby those who are *"Left Behind"* as the book series touts, will believe they are unsaved (though not necessarily true!) People will, as the deception unfolds, think they are facing the GT *without the*

Holy Spirit... since he who *"withholdeth"* was supposedly taken out of the way. Again, it is **not the Holy Spirit,** it is Michael the Archangel who biblically fights with satan, he is withholding him. Doctrine matters! Again, we must each remain focused on our personal forthcoming eternal joy, especially now that it is so near!

We do not, certainly we cannot comprehend the massive horror of the global abortion genocide, from God's vantage point. Nor can we at all relate to the judgments God has enforced on historical people groups, for far less human carnage than our societies have perfected today. "Our day" is overdue it's judgment. Worldwide reaping of judgment, or the enforcement of sowed consequences by earth's Holy Creator, must take place as it always has. Make no mistake, Almighty God is just. Jesus saves sinners, yet no one is favorably entitled to persist in sin above others in history, without retribution.

The Bible is purposefully direct, God is the same *"yesterday, today and forever..."* Heb.13:8 Our Lord will rightly arrange earth's judgment within His balanced scope of all people of human history... Truth is, all bible prophecy will be perfectly fulfilled, whether or not we read, understand, or preach all of the differential parts. Are we in a momentary quiet before the world's longest and worst storm? Is all humanity blind as the prophesied *snare* approaches? The worldly do not realize that they are drenched in guilt, and should be trembling before the start of the judgment of their Holy Righteous Creator. Conversations among Christians today generally sidestep the subject of God's well-defined coming end-time judgment, even though it is more relevant as each day passes. Let us, *"with fear and trembling"* work out our salvation, as per Philippians 2:12-13. Mercifully, in Mt. 24:13, Jesus said,

> *"He that shall endure unto the end, the same shall be saved." We must live in such a way right now, so that God may show-off your and my fearless obedience and trust in Him! With our brother Paul the apostle, we say, "I can do all things through Christ which strengthens me"*
>
> — PLP4:13

4

GOD'S JUDGMENT, TIMING, AND THE BIBLICAL SIGNS

D oes the bible give clues to the start of God's judgment? Fortunately, yes. The Lord loves people, all people, and wants everyone to know details of the judgment coming. Our loving Lord has declared exactly how believers must use their faith... it is to share understanding and prepare themselves, more on that later. Let us study the Bible objectively like the Bereans, so that we might respond as did the wise men of old. Using a Berean, prove-with-scripture approach to establishing doctrine, consider this prophecy Jesus announced, in the book of Matthew chapter 24. Please begin by reading this complete chapter, and have it open beside you as we proceed. It starts with Jesus answering questions about the final *"end of the world"*; keep that in mind. In verses 32-35, Jesus gave a parable that the generation which sees the fig tree bud, meaning Israel bud as a nation, will **not pass away** until all the listed tribulation events occur.

Amazingly in May of 1948, almost two-thousand years after being destroyed, Israel has become a nation again; a unique occurrence in human history. Jesus tells us that **the witnessing generation** which sees Israel "bud" or be reborn, would not die until all listed tribulation events are completed. Reading what Jesus said in verses 34 and 35:

> *"VERILY I SAY UNTO YOU, THIS GENER-*
> *ATION SHALL NOT PASS, TILL ALL*
> *THESE THINGS BE FULFILLED.*
> *HEAVEN AND EARTH SHALL PASS*
> *AWAY, BUT MY WORDS SHALL NOT*
> *PASS AWAY."*

Wow! The universe shall disappear but not Jesus' words! Major topics; let's proceed with definite caution...

The seven year tribulation beginning and full duration until the end of the world, will be completed before the end of a generation. So, how long is a generation? As God would have it, in one place in the bible, a generation is defined for us as being a range, or span of time.

Look at Psalm 90:10. *"The days of our years are threescore years and ten; and if by reason of strength they be fourscore years."* A biblical score is twenty. It will seem extraordinary here and now, to boldly put a pencil to the Bible prophesies. Jesus told us repeatedly to watch for and to be aware of our surroundings. Let us press in therefore, and respect our Lord's proclamation, not with presumption or arrogance, but in devoted earnest obedience of His word.

Applying the verses, a generation is defined as being 70, or if by strength it may be 80 years long. So, the earliest and latest that the seven year tribulation period might end, is within this ten year period. If we add 70 years to Israel's rebirth date of May 1948, we come to May of 2018 - potentially the earliest that the seven year period might have ended. The first half of the seven year period is not indicated to be violent or chaotic. It is not the 'great tribulation'. It has only the rise of a 'White Horse rider' and a secretive Peace Agreement with Israel (Dn. 9:27). President Trump helped Israel with a peace agreement that was signed on September 15, 2020, called the Abrahamic Accords. [Note: this covenant *has all features* of the one prophesied. Web search: "Chrislam: the Abraham Accord, 'Christ End Time Ministries'"] On the other hand, adding 80 years to May of 1948 – is May of 2028. This is the last possible year that the witnessing generation shall not have passed, *"till these things be fulfilled."* So, this date is the farthermost time the great tribulation would end, and the end of the world.

Therefore, if we back up the three and a half year length (from May of 2028) given for the great tribulation period – then the *very latest that the GT can start, is* November 2024. This would mean, that the first quiet half has already begun. Interestingly, November 2024, is the month of the next American presidential election, and also a rare full lunar eclipse. Though, we cannot know the day or hour, vs 36; God has stated that He certainly does want us to know the season of the *second half - the "great tribulation." He has given us a wicked man's name (the white horse rider) to count, and the abomination of desolation signal, and an Israeli Peace treaty, as signs to watch for. It is prophesied to* be 3.5 years long. Again, it will *end between* the 70 to 80 year generation period following Israel becoming a nation- which we witnessed, May of 1948. Respecting our Lord's emphatic statement, *"my words shall not pass away."* Building verse upon literal verse, with the utmost respect for the Word of God, and with trembling awe of God, we must observe, May of 2028 or before, the world will end. One cannot break down what Jesus warned his disciples: *"this generation shall not pass away till all these things be fulfilled."*

So, if we apply Jesus's words from where we are, today is March, 2023, the great tribulation season **may start any moment**, with a possibility being as late November 2024, about 3.5 years before the end of the longer/stronger 80 year generation. We are told this by Jesus, while again the generation range is defined in the Bible by a prayer of Moses in Psalms 90.

In summary, a span of time from God, shows the Lord's great mercy. By using a time range, no man can guess when the day or hour of his return. Jesus knew it would be difficult for us to learn, share, and apply this extreme information. He made his declaration 2000 years ago, with great emphasis, look once again:

> *"VERILY I SAY UNTO YOU, THIS GENER-*
> *ATION SHALL NOT PASS, TILL ALL*
> *THESE THINGS BE FULFILLED.*
> *HEAVEN AND EARTH SHALL PASS*
> *AWAY, BUT MY WORDS SHALL NOT*
> *PASS AWAY."*

We will soon begin to experience the most dramatic evidence of biblical prophetic events - perhaps since Jesus rose from the dead! When the abomination of desolation event signals the start of the great tribulation, Christians will *have a fresh surge of confidence in God's written Word*, ironically, as violence spreads globally out before us. But, juxtaposed in the civilization collapse, will also be an extraordinary temptation to fear. Sadly, to the ignorant masses– confusion and fear will be overwhelming, *"Men's hearts will fail them for fear"*. Lk. 21:26 This season (again, we cannot know the precise start time- day or hour, vs. 36), will begin with the anti-Christ standing in the temple vs.15, and per Daniel 12:11. This single notable event will signal the start of global violence and bloodshed, Mt. 24:22.

Keep in mind, a temple building **does not** need be built for this to be fulfilled, as many teachers claim. A precedent was set in Ezra 3, as sacrifices were made vs.6, at the *"House of God"* vs.8, at a time when ONLY an altar stone was in place; not even the temple foundation was built. Yet workmen were said to work *"in the house of God"* verse 9. Point being, anti-Christ may unwittingly fulfill his prophecy on the current Temple Mount, and stop a Jewish sacrifice, with no temple ever being built. Let's not trip on this issue.

Luke 21:35, *"For as a snare shall it come on ALL them that dwell on the face of the whole earth."* No pre-tribulation rapture… the emphasis is that it comes "on ALL." In fact, who but expectant Christians, waiting for a pretrib early-out, but instead see the GT trauma has begun, would quote the mortified confusion of this prophecy… *"where is the promise of his coming?"* 2 Pet.

3:4-6. Atheists are not looking, neither are Muslims or Buddhists, nor those of any other religion. They are not interested in Jesus, at all. Friend, think about it. This verse is a merciful warning from God. The quote would be made only by expectant believers, who discover they are mistaken about the sequence of Jesus coming back! *If this is you - in simplicity and without anger, but forgiveness toward any leaders, adjust your doctrine quickly*, the pre-trib idea was and is wrong. So, entering the great tribulation, God expects specific things from you and me. We'll be looking at them in detail shortly.

By one statistic, almost fifty percent of American Christians are eager pre-trib believers! Again, that confused prophesied quoted question is written to help you and me! It would only be spoken because believers have not departed, but are facing confirmed great tribulation events. If they don't quickly adjust to Biblical reality- **and they must**, they will likely be totally despondent and confused, feeding on each others fear! For God to have quoted in the bible, the default question that will be asked by misled believers, serves to identify their confusion for their good. **In kindness, God knowingly does so to provide clarity, for believers to quickly toss their mistaken doctrine**. The Holy Spirit hasn't left earth! Believers must adjust at once to face the challenging season ahead. Dear Christian, Do *Not* Fall Away.

The bible proclaims that many who were following Jesus, will abruptly stop and turn away. I pray you and I are not among these pitiful doomed ex-Christians. Some may be people that you and I thought were solid believers. Tragically, they will denounce their salvation and take the damning mark, to eat! Wake up, fellow Grace-soaked Believer! Throughout history, God has allowed generations of Christian followers, to live in caves and catacombs, and face violent persecution; even now in places!

If we compare our culture to persecuted cultures who remained loyal to God, sometimes at the cost of their lives... more than one statistic claims 68% of American men *attending church* and *fifty percent of pastors, view porn monthly!* Repent if this is you! Fast, maybe begin spraying your eyes with pepper spray if you fail. That is mild, Jesus himself said, *"pluck it out"* if your eye offends you! He said, if you look at a woman to lust, you've committed adultery in your heart, and you **will be** *"cast into hell"* (Mt. 5:27-29) Yes, Jesus is coming back for a church without spot or wrinkle. Make no mistake, it will be a church that has no hypocrites!

Bible believers are given instructions to prepare, but not with gold or silver. Stored food is what we are told specifically to provide our households: *"**Blessed is that servant, whom his Lord WHEN HE COMETH shall find**"* *so doing."* Doing what...? *"Watch therefore, for ye know not what hour your Lord doth come. But know this, that if the master of the house had known in what watch the thief would come, he would have watched and would not have suffered his house to be broken into. Therefore **be ye also ready**, for in such an hour as ye think not, the Son of Man cometh. **Who then** is a faithful and wise servant, whom his lord hath made ruler over his household, **to give them meat in due season? Blessed is that servant whom his lord, <u>when he cometh</u>, shall find so doing.** Verily I say unto you, that he shall make him ruler over all his goods."* Mt.24 vs. 42-47).

Heavenly Father, strengthen us- your people, to *"**watch**"* with resolute biblical expectation for the snare coming, as you've commanded. May believers gather food for their families, and thereby more readily avoid the mark of the beast; In Jesus' name, Amen!

Dear fellow believer, establish here and now that you will Stand for Christ in holiness, and provide food for your family, to endure through to the end, no matter what! Hopefully, Jesus *WILL* find faith when he comes. It can happen if you and I will believe every Word that *"proceedeth from the mouth of God"*! Mt.4:4 Peace to you, as you use faith-with-works, watch, share and prepare.

Corrupt Laws and Enforcement, A Great Tribulation Challenge

People rely heavily on their possessions. We also hold dear our reputations. We all naturally recoil at thoughts of entering foreclosure, declaring bankruptcy, having money frozen in a bank, being fired from a position, or in any way the discrediting of our reputation. These are normally major indications of poor judgment, moral failure, catastrophic events, war, etc. With the "Great Reset," the creation, definition and the enforcement of laws will be under evil control. The persecution of Christians will force uncomfortable events like those listed, at some point; the Bible says so.

"Jesus endured the cross, **despising the shame.***"* He persisted through whatever came at him. As God's saints we naturally despise all shame, satanic erosion of our freedoms and personal rights. And like Jesus, we must refuse to comply with their new wicked culture. Doing so however, our journey must not include hatefulness or revenge. Our Lord said to his cruel torturers, *"Father, forgive them."* Can we view *every negative event in the same mindset as did our close brother, the Apostle Paul? He had back whippings, but then expressed* **joy** *for being counted worthy of such persecution.* It is not recorded that he spoke with malice against the harsh rulers or their whipping employees.

Clashing events involving "Law enforcement" and "military" (maybe foreign) will likely increase on the streets. Most men bristle at injustice, and rightly so. However, if a husband/father leaps to involvement to make something 'right' but is killed, he multiplies his family's sorrow and difficulty, and would leave them despondent. Realize, the great tribulation is a fixed prophecy. It is not an era that can be rectified by actions of any person or group. Self-control, prayer, preparation and *"providing for one's own"* family, are the focus of this season, 1 Tim. 5:8. Avoid getting involved in others skirmishes in public areas. We will want to provide a quiet private life. From Bible descriptions, we are certain to see mind-numbing events, live and in color. They may be *real or fake, or a combination.* Let us be prayer warriors for people around us and in so doing, release concerns to God.

Can we open our hand up into the air, to symbolically release losses, various pain, shame and humiliation – without anger, as a sacrificial offering to God? *Tough thoughts.* Our tears are captured in a bottle by God, per the Bible; we're not told why. Who knows, maybe for use in ceremonies in heaven - where we could be honored for our courageous right responses! So let's give God joy - that we are willing to also suffer with Jesus, even if we do so with tears and quivering forced gratefulness. It is an honor, according to the Bible, to be

counted worthy of unfair loss events. Philippians 3:10 ”
That I may know him, and the power of his resurrection, and
*the **fellowship of his sufferings, being made conformable***
unto his death” Everyone in heaven will have their own
death story. We must boldly steer our life, to be worthy
if required, to join them! Of course, the rapture is
coming; a worthy goal is to endure for it.

Remember, the Bible is the sole modifying manual for
our ever-changing mindset, **not** the secular news. In
fact, it might be wise to begin ignoring most news
reports. If a prominent presidential election can be
stolen in America, it is likely that all major news - from
here on, is carefully thought out, created and choreo-
graphed for the overall impact of fear and to manipulate people. Personal peace is certain to be difficult to
maintain if we are feeding on glaring Revelation head-
lines, real or crafted. We might have as our operational
verse Romans 14:17, *“For the kingdom of God is not meat*
and drink; but righteousness, and peace, and joy in the Holy
Ghost.” Monitor the affects of everything you hear and
see, for changes to your mental stability. Here's an
experiment. Read the verses below and notice, before -
and after you read, how the Bible can powerfully
elevate your spiritual, mental and emotional outlook!

It is Written:

> *"For our light affliction, which is but for a moment, works for us a far more exceeding and eternal weight of glory..."*

> — 2 COR. 4:17

> *"Not that I speak in respect of want: for I have learned, in whatever state I am, therewith to be content."*

> — PHIL 4:11

> *"Beloved, do not be surprised at the fiery ordeal among you, which comes upon you for your testing, as though some strange thing were happening to you..."*

> — 1PET. 4:12

> *"if any man suffer as a Christian, let him not be ashamed; but let him glorify God on this behalf."*

> — 1 PET. 4:16

"For I reckon that the sufferings of this present time are not worthy to be compared with the glory which shall be revealed in us."

— ROM. 8:18

"You too be patient; strengthen your hearts, for the coming of the Lord is near." Jam. 5:8 "Rejoice ever more."

— 1 THESS. 5:16

"Suffer hardship with me, as a good soldier of Christ Jesus."

— 2 TIM. 2:3

"But resist him(devil), firm in your faith, knowing that the same experiences of suffering are being accomplished by your brethren who are in the world."

— 1 PET. 5:9

"So they went on their way from the presence of the Council, rejoicing that they had been

considered worthy to suffer shame for His name."

— ACTS 5:41

Having read those few verses, notice how you feel encouraged! God's Word can at any moment, instantly provide a spiritual uplift. Remember to feed on God's Word as crucial food. Feast alone and all together in groups!

Do We Truly Believe Mathew Chapter 24 and 25?

Jesus expressly taught in Mathew chapter 24:1-12, the season, culture and specific events from which will begin the great tribulation. In answer to his disciples question: *"Tell us, when shall these things be? And what shall be the sign of thy coming, and of **the end of the world**?"* Our Lord began with a warning, *"Take heed that no man deceive you."* Jesus told of false Christs and *"many false prophets"* who *"shall deceive many"*, wars will be rumored, famines, and earthquakes occur. He called these the *"beginning of sorrows".*

After physical signs, a spiritual collapse will follow - characterized by hatred and betrayal, *"the love of many shall wax cold".* In this setting, our current grace-soaked western culture must pivot from our dreams and varied pursuits for success and prosperity. Verse 13, describes an immediate switch to fundamental survival as the new goal of Christians- globally: *"But he that shall endure unto the end, the same shall be saved."*

A timing key is then given by Jesus in verse 15: *"When ye therefore shall see the abomination of desolation, spoken of by Daniel the prophet, stand in the holy place, (whoso readeth* [that's us], *let him understand.)"*

Again, Mathew 24:43 and 44, culminate with the crucial need for us to *"watch"* and be *"ready"*for Jesus' return. In verses 45 and 46, Jesus describes just what being ready means - practically. *"Who **then** is a faithful and wise servant, whom his lord hath made ruler over his household, TO GIVE THEM MEAT in due season? Blessed is that servant, whom his lord WHEN HE COMETH shall find so doing."* The whole chapter centers on verse 46. God declares family and church leaders who have stockpiled food, are *"blessed."* Based on the warnings of scripture, they've gathered bulk food provisions for those in their household. Interestingly, Jesus' sermon continues into the next chapter - 25, and expands on the provision theme. Here, he again contrasts thoughtful with foolish responses, *At the time of his coming.*

The parable is of five wise prepared and five foolish unprepared virgins – with the latter ending up begging. They are interactive **believers** waiting with anticipation together for their 'groom,' their savior's imminent arrival. Some are wise and are praised, interestingly, for *not* sharing their limited stored provisions (again, expanding from the previous parable – their limited food supply) with the foolish. History has shown that most people even if they see need coming, simply do not prepare. With global food shortages being strategically created and looming, it will quickly become more difficult to obey these two parables and gather.

Elsewhere, in the book of Revelation we learn a loaf of bread in this setting will cost a day's wage, or about $120, a day's wage being $15 X 8 hours, or $120. Our tendency is to think, 'everything will work itself out fine'. But be aware, historical supply-and-demand days are gone. The coming lack of food is intentional, made to control people! You must know everything is different in America now. Let's face it, we live in a country of 330,000,000 people, who have just watched their presidency stolen in plain sight – and even watched the theft proven publicly, in the national documentary movie, *'2000 Mules'*. A coup took over America, it's citizens were mildly annoyed!

Who cannot now see that foreigners, billionaires, the mainstream media, a false president, along with paid puppet politicians and judges - are steering America toward destruction, into their "Great Reset". Elon Musk had the courage to tweet what we all see, "whoever controls the teleprompter is the real president." You have not had your decision-making processes tested with your belly two or three weeks empty, encircled by equally hungry family. For you to think that you and your group will not take the mark of the beast, without having food, is foolish in the Bible. Begging is the worst default plan.

The Coming Prophesied Temple Sacrifice Needs No Man-Made Building for the anti-Christ to Be Revealed.

A fundamental, but not necessarily biblical push today among some Bible teachers is the idea that a temple building must be constructed in order for the anti-Christ to desolate it. One Scripture quoted to justify this is Revelation 13:6, where it is written: *"And he opened his mouth in blasphemy against God, to blaspheme his name and his tabernacle, and them that dwell in heaven."* For us to accept the idea that a temple building is mandatory for the anti-Christ to be revealed, it must be congruent with the whole of the Bible.

Our tendency is to take a verse, isolate it, and build an idea. This time, instead, let us allow God to define for us His doctrine. We will begin by examining what the Lord has said are His requirements for a correct Jewish sacrifice. His Word says, "*An altar of earth thou shalt make unto me, and shalt sacrifice thereon thy burnt offerings, and thy peace offerings, thy sheep, and thine oxen: in all places where I record my name I will come unto thee, and I will bless thee. And if thou wilt make me an altar of stone, thou shalt not build it of hewn stone: for if thou lift up thy tool upon it, thou hast polluted it*" (Exodus 20:24-25). Here we see, described by God for Moses, what is required for the Lord to accept sacrifices: they are just two things: an altar either of earth or a non-tooled stone, which is put in a specific place, a location, where God has chosen to record His name.

Requirement #1: the Chosen Place for God's Name: Look at the Lord's response later in Scripture, when Solomon built God a house. I Kings 9:3 says,"*I have hallowed this house, which thou hast built, to put my name there for ever; and mine eyes and mine heart shall be there perpetually.*" This temple had been a voluntary idea of King David and built by Solomon (see Psalm 132:1-5 and I Kings 8:12-20). Solomon's temple building no longer exists, yet God said that His "heart shall be there perpetually." We know buildings come and buildings go, but God's "*name*" is there now and "*there for ever,*" with or without a man-made building. It is the place where He has chosen to keep His eyes and His heart and His name.

The following verses repeat that it is a place where God chooses to put His name: Deuteronomy 12:11; 14:23; 15:20; 16:2; 26:2; and I Kings 8:29. Jesus, in Matthew 24:15, reaffirms that a *"Holy Place"* is where the desolation will occur. In these verses, there is no mention of a building. In fact as mentioned, in Ezra Chapter 3, sacrificing is described as being offered at the *"house of the LORD,"* yet only an altar stone was in position. The fulfillment of the desolation prophecy can therefore occur, when a non-tooled altar stone is positioned at God's *place* on Solomon's temple mount, and following a minimum of one day of sacrificing.

The verse could also be understood to mean a sacrifice is not actually accomplished, but the intent to sacrifice could be stopped; thus, the anti-Christ would be stopping the sacrifice. The man who stops Jews from sacrificing will, by so doing, conclusively identify himself as the anti-Christ to the world. This specific day that the sacrifice is stopped is very significant. It will begin the count of 1290 days of great tribulation on earth, according to Daniel 12:11 and Matthew 24:15.

Requirement #2: A Group with a Non-Tooled Stone Seeks to Sacrifice: Today there is a dedicated group of Jews in Jerusalem calling themselves "The Temple Mount Faithful." They have been awaiting the arrival of the Messiah. Not believing in Jesus though, they are awaiting "Mashiach Ben David" or the Messiah, the son of David. They currently have a multi-ton non-hewn stone (and a backup), formed according to the scriptural requirements. Their intention is to place it for an altar and to serve as the cornerstone of the third temple. Their whole desire is to place it on Solomon's Temple mount and then to begin sacrificing. Currently, there is space enough with the Dome intact.

Below are their non-hewn stones which meet the prophecy requirements. The enthusiastic group drives periodically around the Temple Mount and through Jerusalem with these special stones. One attempt to place the stone has failed for fear that the Arabs who currently control the Temple Mount, would resist them. Reading from their website under "Short Term Objectives," they had listed a main goal: "This first stone of the Third Temple will soon be laid." Pictures of the special stone, objectives of the group, and their history are seen at www.templemountfaithful.org

They have announced plans to begin sacrificing on the Temple Mount once a stone is positioned. Again, reading under "Recent Events": "The efforts of The Temple Mount and Land of Israel Faithful movement and others to perform the sacrifice on the Temple

Mount faced opposition of the Israeli authorities, who were scared of the Arab reaction".

They actually had permission and had set a date agreed to by the Israeli authorities, but the authorities called off the event for fear of the Arab response. When they physically place the stone and begin sacrificing or attempt to, they will unwittingly meet part of the details of the prophecy. Such a move however, would likely provoke the fury of the entire Muslim world, since they claim the Dome of the Rock and Solomon's Temple mount are all theirs, and Islamists have shown they would fiercely defend it. In fact, visitors are allowed on the mount, but so zealously is the site watched over, that if anyone moves his mouth without words as if praying or bows his head momentarily as if to worship, he is briskly escorted off the temple mount by armed Islamic soldiers.

This place, the Lord's eternal chosen focal point, is also naturally Satan's focus for subversion. At some moment of time, it will be the pivotal triggering scene for the countdown closure of mankind. One day in July of 2017, Israel took control of the Temple Mount, but gave it right back. The anti-Christ can identify himself once a non-hewn stone is in place and sacrificing is intended or has begun. It may be that he will come acting as a peacemaker or mediator between the Jews and Arabs.

The moment he stops Jews from sacrificing on the Temple Mount begins the 1290 day count of the "*great tribulation,*" as it has been prophesied in Matthew 24:15 and Daniel 12:11. (Stay up to date with the activities of "The Temple Mount Faithful"[1]. To join their mailing list, write to: tmflist@templemountfaithful.org). We have seen that at any time, sacrificing on a qualified altar can occur, at God's chosen place. Everything in that regard is standing by. The only part of the desolation prophecy lacking at that sacrifice, would be the arrival of the anti-Christ, whom our Lord expressly wants us to identify: *"Here is wisdom. Let him that hath understanding count the number of the beast: for it is the number of a man; and his number is Six hundred threescore and six"* (Revelation 13:18)

For you to "understand" and then let others *"that hath understanding count the number of the beast,"* are prophesied directives, defined as a position of Wisdom. At the conclusion of the book of Daniel, he is told specific end-time identifiers. The insights and applications, however, would remain closed by God and sealed until the *"time of the end"*, a time when *"knowledge shall be increased."* The book of Daniel continues: *"But thou, O Daniel, shut up the words, and seal the book, even to the time of the end: many shall run to and fro, and knowledge shall be increased"* (Daniel 12:4).

A Man Who Matches the Features of the anti-Christ.

When I found out about the 666 calculation discovery twenty-three years ago, a friend flew with me from Texas to Oklahoma to meet and visit the discoverer, Monte Judah. During our three-day visit with Pastor Monte and his family, he presented us a copy of a lineage chart showing prince Charles to be a descendant of King David.

[Authors note: ***Dear Christian, DO NOT Fall Away,*** does not claim prince Charles of Wales is the anti-Christ. This book simply obeys the Revelation verse to let you count, and provides some of the unique features that the Bible describes, by which we are to identify a candidate and what we are instructed by God to look for, and to share with others.]

In simple obedience of the *"let him... count"* command, this opportunity to count is prepared for you below. Worldwide positive identification of the anti-Christ is prophesied to be at an *"abomination of desolation"* event in Jerusalem. In fact, there are many additional specifics that currently match prince Charles of Wales with those of the biblical description of the anti-Christ.

Dear Christian, DO NOT Fall Away, has been written in simple devoted obedience to the biblical literal prophetic directive to "*Let*" people "*count*" "*the number of a man.*" For you to share this with others, letting people count, is likewise an opportunity for you to exercise your faith and be obedient to the Bible's straightforward simple prophetic directive. [One way you may "let" others count is to email/post pictures of the cover of this book.]

The matches of his identity will now be demonstrated. 666 Calculated in Hebrew, confirms 666 in English. His name is also his title, and was chosen by him. It equals 666 using the same numbering system that was common when the book of Revelation was written. The King James Bible does not capitalize the title "prince" when speaking of him; therefore in deference to scripture, ***prince*** will be presented in the lower case throughout these pages.

The Hebrew alphabet has a number that corresponds to each letter. This is called the Hebrew Gematria. It was common in the days of John, the inspired writer of Revelation. To grasp this numbering system, we turn to a Jewish website dedicated to its explanation and use: www.inner.org/gematria/gematria.htm [2]:

"*Here is a basic introduction to Gematria that discusses different systems for identifying the numerical equivalence of individual letters, how these letters can be calculated according to the implicit word-value of their letters. The assumption behind this technique is that numerical equivalence is not coincidental. Since the world was created through God's "speech," each letter represents a different creative force."*

Among the applications, the Gematria is used to count the corresponding number of each letter in the name of a person. With the ancient calculation technique installed on a computer, name searches can be rapidly calculated. A Christian Jew, Monte Judah, developed such a program and fed in names from encyclopedias. He made the calculation discovery below. The most universal language of our day is English while the most predominant ancient language was Hebrew. These two bear witness as it is written: *"In the mouth of two or three witnesses shall every word be established"* (2 Corinthians 13:1).

English Alphabet	Hebrew Alphabet	Gematria Value both Alphabets	Name and Title	English Count	English subtotal	Hebrew Count	Hebrew subtotal
A	Alef	1	P	70		Nun =50	
B	Bet	2	R	90		Samech =60	
C	Gimmel	3	I	9		Yod =10	
D	Dalet	4	N	50		Kaf =20	-140
E	Hey	5	C	3			
F	Vav	6	E	5	-227	Tzadik =90	
G	Zayin	7				Resh =200	
H	Chet	8	C	3		Lamed =30	
I	Tet	9	H	8		Samech =60	-380
J	Yod	10	A	1			
K	Kaf	20	R	90		Mem =40	
L	Lamed	30	L	30		Vav =6	
M	Mem	40	E	5		Yod =10	
N	Nun	50	S	100	-237	Lamed =30	
O	Samech	60				Samech =60	-146
P	Ayin	70	O	60			
Q	Pay	80	F	6	-66		
R	Tsadik	90					
S	Kuf	100	W	0			
T	Resh	200	A	1			
U	Shin	300	L	30			
V	Taf	400	E	5			
(W-Z)	n/a	zero value	S	100	-136		
TOTAL					666		666

Unlike Hebrew and Greek, English has no numerals of its own, we use Arabic numerals. Our English alphabet can be assigned the numerical values corresponding to the Hebrew Gematria number placement, with it's 22 letters. Our alphabet with 26 letters, is longer than the Hebrew letter system by 4 letters; therefore, when the Gematria is applied to English, letters w, x, y and z have no Gematria value. All this will become clearer as you obey the prophecy and count *"the number of a man"* for yourself (incidentally, the name Charles means man or manly).

The first two columns above, are the English and Hebrew alphabets. The third column shows the letter value of the Gematria for each alphabet. Each letter in

English and Hebrew represents a number value in the Gematria (third row) of each alphabet. His title is also his name. prince Charles of Wales, is down the fourth row. To count the "number of a man," add the letter values of row 5 for English, and row 7 for Hebrew. This becomes confirmed, as you can count his name in two languages. Lastly, add the subtotal columns of each language to fulfill prophecy, as you count *"the number of a man."* The fact that he has now become king, takes nothing away from the identifying method and purpose that God intended. Yes: *"Here is Wisdom"* - God has lovingly provided a gateway of insight for us to know with certainty, subsequent end-time events are near. We are not intended to grope in the darkness!

"Here is Wisdom," - with our obedience to count and identify an anti-Christ candidate, additional confirming knowledge may now unfold to show vast and final biblical implications. *"Here is Wisdom,"* - our Lord has exposed a conquering ruler early, in order for us to have time to *prepare - like the five wise virgins.* *"Here is Wisdom"* - that in verifying the number of his name, we are able to recognize a modern prophecy, pinpointing exactly where we are along the Bible's overall timeline, prior to *"red horse"* world events.

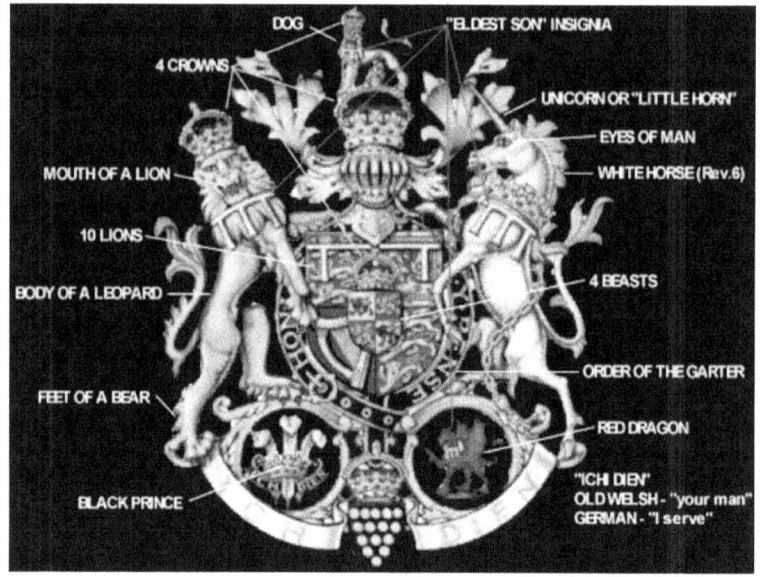

Additional Biblical Traits prince Charles Matches:

The sheer quantity of anti-Christ prophecies he has satisfied, or is currently poised to fulfill, has inspired a number of recent books. We will touch on some of the highlights. Checking Charles' life with other prophecies, the results are astounding. It is written: *"The beast which I saw was like unto a leopard, and his feet were as the feet of a bear, and his mouth as the mouth of a lion: and the dragon gave him his power, and his seat, and great authority"* (Revelation 13:2). This verse conveys additional identifying but cryptic ("like unto") characteristics of the beast/anti-Christ. The scholarly work, "The Antichrist and a Cup of Tea." by Christian author Tim Cohen[3], details some surprising features of prince Charles' coat of arms.

In reference to the pictured left hand, lion-like, guardant dexter (Latin for right - as from the bearer's position) supporter in the prince's coat of arms, Cohen has researched the original meanings of this odd beast, from which the ancestry of the prince is represented: "the leopard-like body is for Germany, the bear-like feet are for France, and the lion-like mouth is for England."

At Charles' televised investiture, the Queen placed the coronet cap upon his head, and led him by the hand to present him to the Welsh people. Watch the full video if you like on YouTube, enter a search for: prince of Wales Investiture at Caernarfon 1969. Oddly, when prince Charles referred to this investiture in 1969, he claims: *"Within the vast ruins of Caernarfon Castle,* **My father** *invested me as prince of Wales. Upon my head,* **he** [see picture of Queen Mum] *put a coronet cap as a token of principality, and into my hand the gold verge of government, and on the middle finger the gold ring of responsibility. Then leading me by the hand through an archway to one of the towers of the battlements,* **he** (emphasis added) *presented me to the people of Wales."*[4]

Queen crowns "prince" Charles on behalf of the dragon

To whom is the prince referring to here as his father, when the queen personally officiated?[5] It was not Prince Phillip; he lacks investiture authority. He is not a British sovereign; besides, Phillip was seated while the coronet cap was placed. For a clue, let us look again to his coat of arms. Everything about the crest holds significance.

Notice the two items floating near the bottom. They are self-supported, and unattached to earthly things; they are spiritual in nature. The floating crown and feathers, display the motto, Ich Dien, meaning, "I Serve," from the German, Ich Diene. The other floating item is a red dragon. The spiritual message conveyed here is clear, *"I serve the Red dragon."* Reinforcing this idea, an intricately carved and prominently displayed dragon identical to the one on the crest was positioned on the Queen's throne during the investiture.

So the Queen had operated in service to, and on behalf of Charles' father, the dragon. The prince of Wales, in his quote above, plainly identifies the dragon as his "father," who has led him. The dragon matches the prophetic identification symbol. Quoting him again: "My *father* invested me as prince of Wales. Upon my head, *he* put a coronet cap as a token of principality, and into my hand the gold verge of government, and on the middle finger the gold ring of responsibility." Now look at Revelation 13:2 again, *"and the dragon gave him his power, and his seat, and great authority."*

It appears Charles himself has lent biblical evidence that matches that of the anti-Christ, proclaiming at his public introduction that the dragon gives him: his "principality," or as the Bible describes it - "***power***"; his "verge of government," biblically called his "***seat***"; and his "responsibility" in the Bible called "***great authority.***"

Anthony Holden reports the investiture highlight. Keep in mind the Queen is acting on behalf of the dragon: "Then came the climax of the ceremony. Kneeling before his mother, the prince of Wales intoned the oath": Placing his hands between the Queen's, he declared: *"I Charles, prince of Wales, do become your* (the dragon's) *liege man* [def: supreme ruler possessing "ultimate power"] *of life and limb and of **earthly worship**, and faith and truth I will bear unto **you** to live and die against all manner of folks."*[6] Charles vowed publicly to receive worship on behalf of the red dragon, or satan. *"All that dwell upon the earth shall worship him, whose names are **not** written in the book of life of the Lamb slain from the foundation of the world"* (Revelation 13:8)

Indeed, the backrest of the queen's throne has the satanic dragon of Wales on it. When she placed the coronet upon her son's head, this dragon was directly behind her and adjacent to prince Charles' throne. Of the anti-Christ, it is written, *"And the dragon gave him his power, his throne, and great authority"* (Revelation 13:2); *"that old serpent, called the Devil, and Satan, which deceiveth the whole world" (12:9).*

As Prince of Wales - prince Charles receives his power, his throne, and great authority from the red dragon of Wales. The command to count for ourselves and also to let others count his name is unquestionably given by the Lord to educate people worldwide, of the vast closing changes that are near.

Summarizing, it is written: *"All that dwell upon the earth shall worship him, whose names are not written in the book of life of the Lamb slain from the foundation of the world"* (Revelation 13:8). The Dragon is displayed on the back of the queen's throne, to whom Charles gave allegiance and for whom he vowed to receive worship. As Christians, we are never to follow satan, or his ways, or his demands, even if it costs us our life. We listen to and obey a different Father – our Loving Creator, and we have eternal JOY awaiting us in heaven - for absolute certain! Meanwhile, the enemies of God, powerful and active in the world today, are just large tumbling boulders careening down a mountain, destroying as they go, hurling towards the lake of fire; where they will remain forever. Rev. 20:15

Salvation in Jesus Christ, is our personal extraordinary treasure!

The Secretive Life of the White Horse Rider:

Most people think of him as just a figurehead. This public anti-hero image belies his thirst for world control. Cohen provides some definition to his power, his seat and great authority, exposing, *"a side to prince Charles, as well as to the British Monarchy as a whole, that has been carefully hidden from the public and most media a side that reveals power, influence, fame, and wealth totally unimagined by the carefully led and spoon-fed public."*[7]

The House of Windsor and prince Charles, perhaps most importantly, have majority control of the world's food distribution network. Preeminent among the food companies is Archer Daniels Midland, otherwise known as the "supermarket to the world." ADM alone has control of 75% of the world's food distribution, effectively possessing the power to decide which countries will eat and which will not. It is written: *"I saw, and behold a white horse: and he that sat on him had a bow; and a crown was given unto him: and he went forth conquering, and to conquer"* (Revelation 6:2).

"Winged prince is Savior of the World"[8] - *LONDON: The prince of Wales is to be immortalized in bronze as a muscular, winged god dressed in nothing more than a loincloth. He will be the first living member of the Royal Family to have a life-size statue dedicated in his honor. Although the prince is*

destined to become 'Defender of the Faith' when he becomes King of England, the inscription on the statue in Brazil will honor him as 'Savior of the World.'" Jose Wilson Sequeira Campos, the Governor of Tocantins, in central Brazil, said: *"It is prince Charles saving the world. We think he is deserving of it."*

Fox News Life, March 7, 2002. NEWS BRIEF:

A brooding mass of humanity at his feet, in what appears to be mud, looks up expectantly at their savior.

The groveling people, one with a wine bottle, repre-senting the world in a mess, are dwarfed by Charles' bigger-than-life size statue. Reading again from the article: *"He is shown naked, apart from the loincloth, with giant, angel-like wings protruding from his back. His arms are extended as if offering comfort and security. The statue will dominate the town square, to be named after the prince, in Palmas, the state capital of Tocantins on the edge of the rainforest. The sculpture, which will invite comparisons with the statue of Christ overlooking Rio de Janeiro from Corcov-ado, is set on a marble base."*

A White Horse Rider Is Prophesied to Rule a Needy World.

About the anti-Christ, we learn: *"He causeth all, both small and great, rich and poor, free and bond, to receive a mark in their right hand, or in their foreheads: and that no man might buy or sell, save he that had the mark, or the name of the beast, or the number of his name"* (Revelation 13:16-17). This prophecy describes an all-encom-passing monetary shift. It is a change that will occur instantaneously by world standards within the 3-1/2 year great tribulation period. It will be a complete substitution of what each nation trusts as their local currency now, to some new system that controls all financial transactions. Such a quick and universal, worldwide monetary change would seem to indicate widespread pre-planning, followed with voluntary submission and implementation. It is as if each country will cry out for this system, perhaps compelled to avoid being left out of international trade, or more likely, to simply retain their ability to buy food and energy.

Following the biblical worldwide *"snare"* [1913 Webster[9] definition: *"to ensnare; to entangle; hence, to bring into unexpected evil, perplexity, or danger."* Cambridge Advanced Learners[10] definition: *"Snare, n., a trick or situation, which deceives you or involves you in some problem of which you were not aware"* (emphasis added)]. It should surprise no sufferer of the financial breakdown how well prepared for and accepted the new beast system will be by governments. World leaders, including all recent American Presidents, have been publicly vocal in speeches referencing an undefined and curiously unquestioned "New World Order," now known as the Great Reset. The following quote from David Rockefeller, founder of the Trilateral Commission, in an address to a meeting of the Trilateral Commission in June of 1991, exposes the extent of this quiet collaboration, decades of preparation, and the groups world reordering goals, along with the media's odd silence on the subject:

"We are grateful to the Washington Post, the New York Times, Time magazine and other great publications whose directors have attended our meetings and respected their promises of discretion for almost 40 years. It would have been impossible for us to develop our plan for the World if we had been subject to the bright lights of publicity during those years. But, the work is now much more sophisticated and prepared to march toward a World government. The supranational sovereignty of an intellectual elite and World bankers is surely preferable to the national auto-determination practiced in past centuries."[11]

No secrecy or shame is holding them back from their goals: *"Some even believe we are a part of a secret cabal working against the best interests of the United States, characterizing my family and me as 'internationalists' and of conspiring with others around the world to build a more integrated global political and economic structure - one world, if you will. If that's the charge, I stand guilty and I am proud of it."* - David Rockefeller from his book, David Rockefeller: Memoirs. Eventual global control is their ultimate goal according to Dr. Johannes Koeppl (Former official of the German Ministry for Defense and advisor to NATO): *"The interests behind the Bush administration, such as the CFR, the Trilateral Commission - founded by Brzezinski for David Rockefeller - and the Bilderberg Group have prepared for and are now moving to implement open world dictatorship."*

The diabolical movement has advanced much today, under a new name, "The Great Reset". Now, brazenly outspoken and confidently backed with cutting edge all-seeing technology, the cabal of billionaires, world leaders, China, the mainstream media, paid-for politicians, and leftist judges – each contribute their part toward the overall goal of "open world dictatorship".

Our Beliefs Pave Our Eternal Destiny.

It is easy to see why an anti-Christ will look, through starving eyes, like a compelling savior indeed. The Bible is unwavering; deception will be extremely convincing. The Beast's answers will be necessary and logical, *for the "good" of all mankind*. While opposition from Christians will likely appear stubbornly archaic, grossly unreasonable, and even blatantly rebellious. The most effective snare will have loyal believers looking like we are to blame, and our rejection, singularly jeopardizing mankind's future. **Except for a few eternity-influencing scripture verses we cling to**, we will not likely have any excuse not to participate. This is the ultimate show down, on a desperate planet, with all systems crashing. Uncooperative believers may be considered treasonous by the spoon-fed mislead masses. This error will become especially explosive if the differences are within families and churches, as depicted in Mark 13:12,13.

Careful and loving communication, referencing scripture, can cancel confusion. Remember,"**Only** *by pride cometh contention: but with the well advised is wisdom"* Prov. *13:10* Be Humble - in reality, only our determined personal commitment to the God of the Holy Bible, particularly with our focus on the joy that God has set before us in heaven, along with rigid abhorrence of the literal, physical, eternal fires of hell, that Jesus so clearly taught about, will be enough to prevent people from standing in line, reaching out their hand to take the cursed mark, and so damn their soul *forever*.

God does not want people ignorant. How dangerous our natural reactionary thinking could be under this stress... "Surely a good God wants me to feed my family, even if it means taking a silly mark - I would still be a Christian, WE HAVE TO EAT!" Friend, you must not doubt this prophecy, it is absolutely clear. THE BIBLE SAYS: *"If any man worship the beast and his image, and receive his mark in his forehead, or in his hand, The same shall drink of the wine of the wrath of God... and he shall be tormented with fire and brimstone in the presence of the holy angels, and in the presence of the Lamb: And the smoke of their torment ascendeth up forever and ever: and they have no rest day nor night, who worship the beast and his image, and whosoever receiveth the mark of his name. Here is the patience of the saints: here are they that keep the commandments of God, and the faith of Jesus. And I heard a voice from heaven saying unto me, Write, Blessed are the dead which die in the Lord from henceforth: Yea, saith the Spirit, that they may rest from their labors; and their works do follow them"* (Revelation 14:9-13).

Consider also, II Thessalonians 2:9-12:*"Even him (anti-Christ), whose coming is after the working of Satan with **all power and signs and lying wonders, And with all deceivableness of unrighteousness in them that perish; because they received not the love of the truth,** that they might be saved. And for this cause God shall send them strong delusion, that they should believe a lie: That they all might be damned who believed not the truth, but had pleasure in unrighteousness."* These verses explain how that widespread deception will snare non-believers and the ungodly. Again, we must avoid at all costs, taking the eternal-damning mark of the beast. Our singular goal is to joyfully enter into heaven. There will not be one person in heaven who has the mark of the beast. If you find out this information late, and you have taken the mark already, Bible verses speak of cutting off a body part – hand, that offends you. Can you physically remove the mark somehow out of your body? It is surely worth every effort...

"The people that do know their God shall be strong and do exploits. And they that understand among the people shall instruct many"

Daniel's prophecy says: *"They shall place the abomination that maketh desolate. And such as do wickedly against the covenant shall be corrupt by flatteries: but the people that do know their God shall be strong and do exploits. And they that understand among the people shall instruct many"* (Daniel 11:31-33). So, how will you creatively perform our Lord's new marching orders to: *"Be strong," "do exploits"* and *"instruct many"!* God is replacing our age-of-Grace focus - the pursuit of dreams, seeking wealth, and enjoying life's comforts. He wants nimble Christians to adjust quickly to basic survival, share information and prepare. We are to help others survive as well. *"he that shall endure unto the end, the same shall be saved."* Mt. 24:13.

This book can help in that regard. You have just learned about people, places and conditions, that by all appearance, represent the details required for the abomination of desolation signaling event that is prophesied. You know these things because a key verse instructing us to count *"the number of his name,"* has been discovered and boldly shared.

These appear to be the components of an approaching strategic warning sign that God Himself has selected to announce to the world the start of a final period of great tribulation. Specifically, you have learned of a group of Jewish people, who are anticipating their Messiah, a son of David. The Jewish group possesses a biblically mandated non-hewn stone. They have as their sole intention to place that stone on the temple mount, at the single "*place*" on the planet that God has chosen to put His name "*for ever.*" Incredibly, they state their purpose is to begin Old Testament sacrificing. Finally, you have learned of a man with the detailed pedigree described in Scripture: a prince whose name equals 666.

These highlights are readily verifiable, and would complete the stage for "*the abomination that maketh desolate*" event, spoken about in the Bible. Your faith must not be based in a religious institution, or any man's opinion of prophecy. But instead as in Jesus' day, the safest understanding is to limit our hearing to the Word of God, then simply compare events and people that God said would come, to the activity of our day.

A good example of some people giving a correct response to newly presented prophetic information is found in Acts 17:11-12 "*These [Bereans] were more noble than those in Thessalonica, in that they received the word with all readiness of mind, and searched the scriptures daily, whether those things were so.*" Let us be like the Bereans.

Since this information appears to be the alignment of the most pivotally important closing prophecies facing the world, they are worth eagerly examining, sharing, and discussing - *Berean style.* If you have read, studied, and you now recognize these things, you too *"understand among the people,"* as mentioned in Daniel 11:33. An action is expected in response; that is to *"instruct many."* Neither you nor I can know on what day all this will begin, exactly what it will look like, nor should we guess. Emailing and posting copies of this book cover can notify people.

Our challenge of faith is to obediently count his number, study the developing indications, and respond with wisdom by sharing these insights, or as the verse says, *"to instruct many."* Finally, we are to be watchful and prayerful and to take heed to ourselves. In other words, respond by preparing; more verses on readiness will be discussed in the last section of the book.

While the *"falling away"* will tragically include some of today's *respectable* 'Christians', the multitude who obey and *"understand and instruct many,"* may very well be people whose lives are chaotic now with energetic sin, but who look up and recognize an unprecedented opportunity to repent, pray, accept the free salvation gift in Jesus Christ. These will be joyful last minute conversions; people that turn from their sinful lifestyles, who sprint into Jesus' loving embrace of forgiveness. These will triumph at the end, and have great rewards on judgment day!

A heartbreaking transition of **many**, again, who consider themselves decent Christians, are prophesied to change course and become fallen eternal failures; do not be influenced by these. Also, be aware, among those who end up falling away, will likely be renowned and respected prominent 'Christians' who vocalize opposition brazenly, against the obvious biblical signals. Just remember, any person may accept Jesus Christ, embrace these historical closing moments, evangelize, and end up brilliant faithful successes, entering heaven, following this season of judgment.

"Faith Without Works Is Dead."

In the great tribulation, every person will face a compulsory choice to worship either God or Satan (ostensibly the anti-Christ). Loyal believers living at this time will be the Lord's occupying soldiers. We must not question the judgment of our eternal Supreme Commander. Please be aware, that it was the religious leaders of Jesus' day, confident they understood God and His agenda best, who most zealously distrusted Jesus. They faltered at His humble carpenters appearance and revolutionary teachings. While guardians of

God's Holy Scriptures for the world, many of these leaders failed to rightly apply them and lead others, as they should have.

They only had limited doctrinal facts available to bridge the chasm from godly obedience of the Law - which they were commanded to do, over into Jesus' new scandalous idea of salvation, as a work-free universal gift of grace. In a similar way today, we Christians have pushed grace to outer extremes, as boundaries of sin are blurred, and blatant sin even thrives in some churches.

Certainly, our modern New Testament doctrines have especially well emphasized the cheerful benefits of grace. We too are vulnerable however, with scant doctrinal bridging in place, to transition today's believers into a rugged survival lifestyle to endure, in a world snared in satan's wrath and God's judgment. Most people (including me) have little awareness about how to suffer or face death for Christ. Moreover, how will our godly leaders respond to live-in-color events, if they are not arranged as their years of publicly proclaiming Jesus' Second Coming?

Let us pray the Lord's lead spokespersons today, now without their bank accounts and jet fuel, will be responsive to His Word and will make the doctrinal mega-shift to go from emphasizing the many blessings of God's grace, to speaking about the new doctrinal relevancy of enduring all manner of suffering, in the sufficiency of grace.

> *"Beloved, think it not strange concerning the fiery trial which is to try you, as though some strange thing happened unto you: but rejoice, inasmuch as ye are partakers of Christs sufferings; that, when his glory shall be revealed, ye may be glad also with exceeding joy. If ye be reproached for the name of Christ, happy are ye; for the spirit of glory and of God resteth upon you: on their part he is evil spoken of, but on your part he is glorified... Yet if any man suffer as a Christian, let him not be ashamed; but let him glorify God on this behalf. For the time is come that judgment must begin at the house of God: and if it first begin at us, what shall the end be of them that obey not the gospel of God? And if the righteous scarcely be saved, where shall the ungodly and the sinner appear? Wherefore let them that suffer according to the will of God commit the keeping of their souls to him in well doing, as unto a faithful Creator."*

— (I PETER 4)

"I am ready not only to be bound but to die for the name of The Lord Jesus."

— (ACTS 21:13) (SEE ALSO II CORINTHIANS 1:5-7 AND 12:9-10)

Perhaps the greatest qualities each person will need as things shift, is the simple ability to release dreams, drop partially accomplished plans, relearn, prepare for, and adjust how to live content, with only basic daily needs being met. Making such vast life changing adjustments, while keeping communication and emotions within loving boundaries with those we care for – will certainly require the biblical fruits of the Holy Spirit to dominate: love, joy, peace, long-suffering, kindness, goodness, faithfulness, gentleness, and temperance [self control] (Galatians 5:22-23).

Imagine an ideal leader, a man/woman who is out front, industriously preparing, while loving everyone around them with dignity and good humor! Tough times can be faced with a trusting joyous heart of faith, or with prideful contention, it is a choice. We will close this section with some additional supportive biblical evidence for the end of the world.

The largest nation of the world, China, now has an ability never seen before. They can call out an army of

200 million soldiers as prophesied in Revelation 9:15-16: *"The four angels were loosed, which were prepared for an hour, and a day, and a month, and a year, for to slay the third part of men. And the number of the army of the horsemen were two hundred thousand thousand: and I heard the number of them."*

Such an army is defined, and their planet-burning prophecy is given in the book of Joel, chapter 2. Let us take a moment to study this second chapter in Joel. It states that an army's fire-starting activities will be the cause of the sun and moon darkening. Pay special attention to the verses that confirm an end-time context. Joel 2:2 *"a great people and a strong; there hath not been ever the like, neither shall be any more after it"* (Revelation 9:15-16). Also note Joel 2:3: *"A fire devoureth before them; and behind them a flame burneth: the land is as the garden of Eden before them, and behind them a desolate wilderness; yea, and nothing shall escape them"* (refer to 2 Peter 3:7); Joel 2:10: *"the sun and moon shall be dark, and the stars shall withdraw their shining"*(see Revelation 6:12-13, Matthew 24:29, and Mark 13:22-27). Joel 2:11: *"the day of the LORD is great and very terrible"* (note Acts 2:20). Lastly, Joel 2:31: *"The sun shall be turned into darkness, and the moon into blood, before the great and the terrible day of the Lord come."* (This verse is referred to and actually quoted in Acts 2:16- 20. Take a few minutes to read the whole book of Joel, with this context in mind.

God does not want us ignorant of the future, but to be prepared, not in the dark. With that said, let us look at some additional closing indications in the book of Daniel. In the seventh chapter, Daniel has seen a vision of four beasts, which, as he describes, made him *"grieved in my spirit"* (Daniel 7:15). Afterwards, *"one like the Son of man"* (7:13) appeared. A heavenly messenger came to Daniel, who when asked the significance of the vision, gave Daniel the interpretation:

"The fourth beast shall be the fourth [the last] kingdom upon earth, which shall be diverse from all kingdoms, and shall devour the whole earth, and shall tread it down, and break it in pieces (Daniel 7:23). And he [anti-Christ] shall speak great words against the Most High,and shall wear out the saints of the Most High, and think to change times and laws: and they shall be given into his hand until a time and times and the dividing of time (Daniel 7:25, also see Revelation 13:5)

Elsewhere in the book of Daniel, Nebuchadnezzar is given a dream by God, but he was unable to remember it (read the story in Daniel 2:1-45). Daniel then described to the king his dream, tells him that it came from God, and goes on to tell each detail of its meaning. His dream was of a tall statue of a man made in stacked layers of different metals. Each metallic layer represented a separate kingdom period of human history.

The very last era he described are represented by the feet, made of an odd combination of iron and clay. This is an unmixable blend, which represents the last strong and weak nations that have merged. They show the final end, as they are smashed and scattered by the Lord himself. This ending of destruction is a description of all human kingdoms and all human history, being removed forever:

> *"In the days of these kings shall the God of heaven set up a kingdom, which shall never be destroyed: and the kingdom shall not be left to other people, but it shall break in pieces and consume all these kingdoms, and it shall stand for ever."*
>
> — DANIEL 2:44

It is at this time that God sets up His eternal kingdom. The current awkward strong (Germany) and weak (Italy, Spain, Greece) European nations, now merged under the teetering Euro currency, fit the iron/clay likeness well. They appear to correlate with the biblical reference of the last geopolitical arrangement before the end, when Jesus comes and sets up His everlasting kingdom.

God Commands Us to "be not ignorant"

The time of man's final judgment and closure, is also revealed in a double meaning given in II Peter 3:7-8: *"The heavens and the earth, which are now, by the same word* [spoken to create them*] are kept in store, reserved unto fire against the day of judgment and perdition of ungodly men. But beloved, **be not ignorant of this one thing, that one day is with the Lord as a thousand years, and a thousand years as one day."* This verse gives us a command: We are to *"be not ignorant of this one thing"* that a thousand of our years is as a single day to God. Let's examine what that means, so we're not ignorant.

Each of the six original days of creation, and separately the seventh day of rest, also correspond to one thousand years of human time. This alternate meaning of day is applied by God, surprisingly, in the third chapter from the beginning, **and** the third chapter from the end of the Bible! Notice, the doctrine is presented symmetrically within the Bible, providing additional supernatural confirmation.

Look first in Genesis chapter 3. Here, the Lord displays a use of His thousand-year "day" when He enforces

Adam's guilty verdict. Adam and Eve were instructed in Genesis 2:17, "for *__in the day__ that thou eatest thereof thou shalt surely die."* Adam ate the forbidden fruit vs 3:6, and was cursed and expelled from the garden that day. However, God said Adam would die *"in the day"* that he ate it. We know Adam didn't physically die until centuries later, at 930 years old. So God's warning, that Adam would die *"in the day,"* meant within a God-sized thousand-year day.

Now let's go to the third chapter from the end of the Bible. We understand that after the sixth day of creation, the Lord entered a seventh day of rest. Looking at Revelation 20:4 below, (see also verses 5-7), an interesting parallel is given in relation to the original seventh day of rest. Notice, God has planned a joyful one-thousand year period to immediately follow the fiery calamitous close of earth's six thousand years. Look how He even expresses the timing to be right after the great tribulation, as people who gave their lives to avoid the mark of the beast, are celebrated in the scene... verse 4:

"I saw thrones, and they sat upon them, and judgment was given unto them: and I saw the souls of them that were beheaded for the witness of Jesus, and for the word of God, and which had not worshiped the beast, neither his image, neither had received his mark upon their foreheads, or in

their hands; **and they lived and reigned with Christ a thousand years.***"* Thus, in the (blessed) judgment of the saints, are people present who have avoided the beast's mark at the cost of their lives.

In summary, we have just obeyed 2 Pet. 3:7,8 and avoided being *"ignorant of this one thing"*; God-sized days are to inspire us! We are encouraged that a planned one-thousand year era in heaven begins immediately after the great tribulation! We do NOT want to miss this fantastic millennial day of restful fellowship with Christ and our loved ones. This last *"day"* corresponds to that original seventh creation day of rest. From these verses, the Lord shows that a day for Himself also defines a one thousand year era for humanity. He even reveals this with divine symmetry equally, at both ends of the Bible. So, in applying a millennial *"day"* in which Adam receives his punishment, and a closing, thousand-year rest "day" at the end of Revelation, God provides for us to *"be not ignorant"* of our future!

Specifying elsewhere when the end of the world will come, Matthew 24:14 gives another clear signpost. It is written:

"This gospel of the kingdom shall be preached in all the world for a witness [a sign] unto all nations; and then shall the end come." As you read this sentence, Christian TV broadcaster, TBN, has strategic satellites providing 100%

global signal coverage, they say, *"for a witness unto all nations."* We therefore, can expect the end will now come. *"Trust in the Lord with all thine heart and lean not unto thine own understanding. In all thy ways acknowledge him, and he will direct thy paths"* (Proverbs 3:5-6). We are to completely trust in God, not over-think things, talk fondly about the Lord, walk by simple faith; as we do so, He **will** direct our paths!

5

WHAT ON EARTH SHOULD BE OUR RESPONSE?

Though we are God's purchased possessions, like the Israelites, we must walk through some plagues and pass some tests of our faith, as they did. Tragically, they failed later, by responding fearfully when at the threshold of their Promised Land. Those ex-slaves had fine tuned their lives to avoid conflict with authorities. Once free, God took good care of them; He provided all their meals, a warm fire by night and shady clouds by day. In their new comfort, they became smug. They soon despised their new life of faith. God tested them and required they switch from comfort and routine, to begin exercising city conquering courage, and trust.

Our lives are fine-tuned to maintain routine comfort and security. Similarly, we Christians could flippantly embrace similar disdain for a new multi-year change to a retro-pioneer lifestyle. Doing so, we could face failure too, only eternally worse if we listen to the wrong voices, as they did. The intake of our eyes and ears must be supervised like never before. Remember, news from unbelievers caused all but two men, of millions of people, not to enter their Promised Land. All twelve spies saw the same things, but only two decided that God was desirous and capable of bringing them into their home-land. With all of the miracles the Lord had shared with them, this had to be deeply disappointing to the Father.

Today, we simply trust that the Lord is working all things together for good for us, whatever lifestyle changes we must make.

Regulating Our Ear-Gate And Eye-Gate: The Bible Says, Are Important At This Time

Again, whom we choose to listen to during the upheaval, the television and media, misled Christians speaking contrary to Scripture, or... *every written word of the Living God,* is who we will be leaning on to help form our thoughts, regulate our emotions, and influence our decisions. **I sense the need to stress this.** We are arrogant if we think we are more mentally and emotionally resilient than the escaped Israelis were.

Can we listen to lies or misguiding news from media giants ranting in unison against God's purpose for our lives, and honestly think it will not affect us? Faith comes by hearing God's Word, doubt certainly can come by hearing erroneous words, too!

Remember how the Israelis, as a whole, allowed their hearing to instantly flood in doubt about God's promises. We can be infected quickly likewise, if we let the media direct our lives, using fear-filled reports of the devastation, to manipulate us with Bible-opposing instructions. Abraham did not follow the angels to Sodom and pull up a chair to watch God's exciting judgment activity. Nor do we see an account of him poking through the rubble later, or even looking for Lot, his own relative. He spoke with the angels, and then continued to mind his own business; he did not gawk at the judged unfortunate ones.

We don't need to watch the turmoil details either. Let us learn a lesson from Lot's wife, who perished as she turned to watch God's judgment falling. Posting or sending a pictures of this book and sharing this information as an early warning to your family, relatives, neighbors and circle of influence, may limit your feeling of need to check on them later, as well. Perhaps our greatest challenge will be monitoring what we listen to and gaze upon, the airwaves remember, will be filled with lies:

'Christ is in here' or, 'He is over there'. DO NOT BE DECEIVED!

God Tells Us Where to, and Not to Have Our Focus During the Tribulation

Consider the repetitive visual characteristics, in **these seven tribulation verses in a row.** I am making the important point that Biblical signals are what we must look for rigorously and study. The airwaves remember, will be packed with lies and totally convincing deception. As noted earlier, artificial intelligence software can make **anything** look entirely real!

Luke 21: Vs. 25-30:

- "And there **shall be signs** in the sun, and in the moon, and the stars."
- "Men's hearts failing them for fear, and **for looking after** those things which are coming on the earth."
- "And then **shall they see** the Son of man coming."
- "And when these things begin to come to pass, **then look up**, and lift up your head"

- "And he spake to them a parable; **behold** the fig tree."
- "When they now shoot forth, **ye see and know** of your own selves"
- "So likewise ye, **when ye see** these things come to pass, know ye that the kingdom of God is nigh at hand."

Matthew 24:

- Vs. 15 "When ye therefore **shall see** the Abomination of Desolation spoken of by Daniel the prophet. "
- Vs. 33 "when **ye shall see** all these things, know that it is near, even at the doors."

You can go back and read the full verses; my purpose here is to emphasize the 'visual' repetition in this string of verses. Courageous people being observant and watchful, in short, looking for and witnessing the signs of the glorious return of our Lord and Savior Jesus Christ, must characterize the use of our visual energy. With those seven verses in a row imploring us to compare our surrounding sights to Scripture during these times, it is likely the temptation to be deceived and misdirected will be very, very powerful.

The single verse, Luke 21:26 above, gives an extra caution. It describes a specific danger in some *"looking,"* particularly for people who dwell on the tragic 'after-news'. The verse warns us of heart failure from unchecked fear from *"looking after"* those things which are coming on the earth. It is interesting that looking or watching scenes afterwards, is something that is possible if captured on video to be seen later, which has only been available in recent times. Likely, the media outlets will offer heart-rending interviews with the grieving, along with fear rousing views-from-every-angle video worldwide, after tragedies the Bible ascribes to Satan, occur (Revelation 12:12). We might expect a wooing of the masses with strong propaganda and threats, while using very sound, logical, and obvious reasoning, combined with seductive flattery. Satan is the father of lies; he specializes in using charismatic leaders, full of altruistic motives and ideas, that sound wonderful and will likely be absolutely essential for humanity to continue. But their plans will prove to be unbiblical. God's command to us is simple. We are not to feed on hype from the media, by *"looking after"* the devastating events.

Understand, these media people will be clueless, and very confused. They are bold spokespersons, who while disregarding all Scripture, endeavor to define a wrath inflicted by a devil they consider mythical, while having rejected their loving Creator, who has allowed the devil's evil actions.

If you are working in public media, are a part of compromised politics, or are among the world's wealthy - and you are involved in supporting this scandalous power-play against God's beautiful earth and people made in his image, STOP! REPENT! Yes, turn your life to the Lord Jesus, and call out to Him; you may instantly be born again this moment! See page 38. The fearless in the faith, who have understanding and perhaps provisions, and who can hold their lives lightly, like the apostles, will have the opportunity to win souls like none in history! I envision Paul and Peter talking in heaven, as they look toward you and your family when these trials are over. Leaning closer, I can hear Peter timidly saying, "There are some of those courageous tribulation saints; lets go meet them!

"As it was in the days of [Noah], so shall it be also in the days of the Son of man. They did eat, they drank, they married wives, they were given in marriage, until the day that [Noah] entered into the ark, and the flood came, and destroyed them all" (Luke 17:26-27). The point of including this verse, is that Preacher Noah warned those people that a flood was coming. *"By faith Noah, being warned of God of things not seen as yet, moved with fear, **prepared** an ark to the saving of his house"* (Hebrews 11:7). Notice the key word

is prepared. This word defines my second purpose for writing this book (the first is to obey and let believers count the number of his name).

The Bible doesn't say how long Noah pondered God's warning, but once he was convinced that He was serious, he took action and rallied his family together. They set aside their dreams, made lifestyle changes, redirected assets, and together put all their effort into an outrageous activity, preparing a huge boat for a worldwide flood - when it had never even rained on earth up to that point, according to the Bible. Noah's countrymen heard his warnings, but decided to wait to see what would happen. They ended up experiencing God's judgment from outside the ark. It was only Noah's responsiveness to God's Word, to prepare, that separated him from everyone else in his world.

Our Current Lifestyle Should Include a New Activity.

Like Noah, we only have God's Word about this earth's cataclysmic closure. The great tribulation, signals and descriptions, are not superfluous bible filler words to be ignored, but neither should they cause paralyzing fear, what then? As the famine of Joseph's day had early warnings, God wants His people to recognize the times, seek Him, prepare, and endure to the end. Unlike

Joseph's day, there is no one who will be enforcing a food-gathering regimen on anyone. You will be either as one set of five virgins - supplied, or as the other group - begging.

It seems quite clear that God is presenting current Christians an occasion to demonstrate that our modern religion, founded in the Bible, is a relevant responsive relationship, not merely an enjoyable social habit or a pretense for business networking, etc. When to start food and equipment gathering plans and activity will come only from your personal faith that such preparations must be made. Benjamin Franklin states, "by failing to prepare, you are preparing to fail." God has always meant what He says. He will again, mean every prophesied word. With this in mind, there is one thing every believer will agree with; it is the clear obligation we each have to take care of our family, with all of our ability, even to our last breath. To do otherwise is a grievous sin: *"If any provide not for his own, and specially for those of his own house, he hath denied the faith, and is worse than an infidel"* (1 Timothy 5:8). Such a declaration, counter-weighted by God rejecting us if we don't attempt this responsibility, implies that providing is something He knows we are capable of accomplishing. The Lord loves to see even an inkling of faith; He rushes in and shows Himself strong.

Planning to Plan...

The Bible says hunger and thirst will be a central challenge for people in the tribulation era. It is written: *"These are they which came out of great tribulation... They shall hunger no more, neither thirst any more"* (Revelation 7:14, 16). History has shown that every town, city, and country is more or less nine missing meals away from anarchy, even when electricity and water are still available. Plans within a family, small group, or church to *"endure to the end"* must be constructed on scriptural principles. Considerations may include where to live, keeping in mind it may be without utilities. The New York City blackouts and looting remind us of the nature and speed with which the shameful in an urban culture violently react in outages.

The idea of not having the ability to buy from any store anywhere is hard to comprehend. Imagine if every food source wholesale and retail outlet, fast-food and restaurant location you know, suddenly had their doorway permanently policed from you entering; where would you turn? Buying and selling may be abounding at stores all around, though perhaps primarily in the cities, where food and electric power will likely to be guarded. Control over all financial transactions seems to indicate use of computers, needing electricity. Faithful Christians will not have the required mark on their forehead or hand to purchase anything from the new satanic system.

The Bible teaches that all people of the world will be extremely hostile toward Christians in the great tribulation (see Luke 21:16-17). Generally, a political power grab is most successful when it is swift and ruthless and creates intimidation and fear. Satan knows a central threat to his global control, is for mark users to quietly provide provisions, in a sympathetic way, to Christians. Expect immediately, vehement lies against Christians on the internet and airwaves to begin. I imagine this will boost public hatred for and even the accusing of believers.

The scheme worked against German Jews effectively. Those people, quietly going about their business, were suddenly demonized. According to Hitler, "All propaganda has to be popular and has to accommodate itself to the comprehension of the least intelligent of those whom it seeks to reach." The public, in general, is shockingly easy to completely mislead; much more, if all media voices become unanimous, opposing uncooperative, irrational, (fill-in-the-blank) Christians. Rejecting the mark may mean you are refusing to work and pay debts. On the lighter side, if we go up live in the rapture at the end of the world, and avoid taking the mark, this is a time when neither death nor taxes are certain! This brief anti-Christ government is satanic and zealous against God and His people. Like Shadrach, Meshach, and Abednego in the book of Daniel, we too could face a courageous choice to embrace salvation at a high cost. True Christians are ready to release everything valuable on this earth, to keep their infinitely valuable eternal life. We're in good company, filled with God's Holy Spirit, linked with other believers millions alive around the world, who will refuse to worship the beast by taking his damning mark.

An Undefined New Testament Parable: of Christian Selfishness Explains A Coming Need.

A seemingly unchristian parable, describes for us, the stresses of sharing, or rather, wisely not sharing, expressly at "the end of the world."" The Bible has a peculiar parable: A waiting wedding party with some who are wisely prepared, who callously reject the pleas of their needy waiting friends. It turns out that the meaning of this New Testament parable is intended for, and targets a coming need in the great tribulation. Let's study the parable.

What can be done now to avoid the tragic lines of people worldwide, who will stand eagerly waiting to take the damning mark to continue eating and be a consumer? More specifically, what does God say will be a vital activity Christians should be doing, at the point in time He arrives? Answers to both questions are given as Jesus responds to His disciples questions: *"Tell us, when shall these things be? And what shall be the sign of thy coming, and **of the end of the world?**"* (Matthew 24:3). Near the end of Jesus's answer (Mt. 25:1), our Lord gives a key parable of ten wedding virgins, of which five were wise and prepared, and five were foolish and did not prepare: *"The foolish said unto the wise, Give us of your oil; for our lamps are gone out. But the wise answered, saying, Not so; lest there be not enough for us and you: but go ye rather to them that sell, and buy for yourselves"* (Matthew 25:8-9). This puzzling parable highlights a time when God considers people wise for steadfastly not sharing their limited provisions. The foolish are the ones without; they made no preparation and had to beg during a dark waiting season anticipating their lord to arrive. Being called "foolish" indicates they could have gathered provisions, but neglected to do so.

The final challenge given by the wise to the foolish, to go and do what would normally have been easy and obvious, to *"buy for yourselves"*, is the distressed conclusion to the dialog in the parable.

The meaning of the parable - to store food provisions, becomes clear as we read this same parable in Luke:

> *"Let your loins be girded about, and your lights burning; and ye yourselves like unto men that wait for their lord, when he will return from the wedding; that when he cometh and knocketh, they may open unto him immediately"."Be ye therefore ready also: for the Son of man cometh at an hour when ye think not. Who then is that faithful and wise steward, whom his lord shall make ruler over his household, **to give them their portion of meat in due season?** Blessed is that servant, whom his lord when he cometh shall find so doing"*
>
> — (LUKE 12:35-36, 40-43)

Notice first, the verses describe the parable: *your loins girded... your lights burning (store of oil)... like unto men that wait for their lord,* are images used to define needs at Jesus's Second Coming, along with a command, to be ye ready: *"Be ye therefore ready also: for the Son of Man cometh at an hour when ye think not"* Being ready, according to the parable, is having a stored supply- **not** of literal oil in the parable, but food portions for your household, in a needy dark waiting season. Look again. What single activity does the Lord build toward and specify that a *"blessed servant"* should be *"so doing"* at the point in time *"when he cometh"*? He is to be wisely providing portions of *"meat"* (food) for his household, during a prepared for dark waiting season. The meat in the verse is not spiritual food, as some claim; 'spiritual food' is never portioned. God always wants Christians to generously share all spiritual insights; besides, spiritual food is free, and they were told to go buy.

The Bible is talking about literal physical food; this is what is shared in limited portions (or withheld). In the dark tribulation waiting season, as Revelation warns, food will be limited and very expensive. Servings of food or portions at that time, will need to be given out from a storage stockpile, in measured amounts to have enough to last the full duration. This is a whole *"season"* worth of food to feed your complete *"household"* until *"He cometh"*. *"Who then is that faithful and wise steward, whom his lord shall make ruler over his household, to give them their portion of meat in due season? Blessed is that servant, whom his lord when he cometh shall find so doing."*

So in the unmistakable context and timing of this parable, Jesus is answering, *"when shall these things be, And what shall be the sign of thy coming, and of the end of the world?"* These same answers of Jesus are seen also in

Matthew 24:45-47 AMP: *"Who then is the faithful and wise servant whom his master has put in charge of his household to give the others [in the house] their food and supplies at the proper time? Blessed is that [faithful] servant when his master returns and finds him doing so. I assure you and most solemnly say to you that he will put him in charge of all his possessions.*

In light of this new and challenging perspective, I urge you to take a moment to read Matthew, chapters 24 and 25; it is all one sermon with mostly this single theme. The Bible is very clear; the wise survive while waiting for Jesus to come back, at the end of the world, by making preparation with stored food; in other words, a rationing program. Foolish people, like those unresponsive and unprepared in Noah's day and the unwise wedding virgins will think they will just get by, because the government always helps, and it will come through again. This mindset and people now living off the government may in their dependence need the strongest personal determination to avoid the mark of eternal destruction.

The US population, remember, has lined up for meals once before in the 1929 stock market crash. Millions of Americans were starving. Collectively nationwide, many miles of people lined up in cities daily, helplessly waiting for free government food without which they would have died. This same pattern of desperation is shown in Genesis 41:55, where people cry out to Pharaoh – their government, for bread. In the future, we are told, it will cost a person his eternal life to participate. To buy food in the new economy, a person will have to have a damning mark of the beast.

Finally, it is not faith but foolishness to expect manna from heaven again; we are not rushed slaves leaving another country in one day, like the Israelis, without ability to prepare. Frankly, to assume you can avoid taking the mark of the beast, without having at least made attempts to prepare, may be presumptuous.

None of us have performed a multiplication of food even once, much less for several years. Neither have we had our thought and decision processes tested, with our stomachs two or three weeks empty, surrounded by equally hungry friends and family members. (See Deuteronomy 28:53-57) DO NOT BE AMONG THE WAITING, NEEDY, BEGGING FOOLISH. Please read the parable of the ten virgins again; the need to gather food rations or portions to endure the tribulation is the whole point of the wedding virgins parable-turned-prophecy (and thus a major take-away point of this book).

I offer the following true story to illustrate the creative and assertive urge people have to avoid starvation. One man, who was a prisoner in a Nazi concentration camp, told how to survive, he sifted through prison guard's stools for potato peels to rinse and eat. Americans are so familiar with food on every hand, yet we are not aware just how vulnerable to being without we really are. In the other extreme, that of self-restraint, people have endured a self-imposed hunger strike, with rigid self control, and avoided eating food presented to them, until they passed on.

We can go without food, if our eternal life depends on it. Taking the mark is a greater sin than that of Esau, who gave his earthly birthright for food (Genesis 25:32-34). It is worse because a tribulation sufferer, who turns away and takes the mark, will be terminating his eternal birthright and bringing the judgment of hellfire upon himself, similarly to Esau, for momentary food.

Doubtless, many neighbors, friends, and relatives will not know to be prepared. *Love* prompts us to warn people; *Wisdom* dictates it be done early; while *Faith* makes preparation for as many as possible, starting with those of your household and then expanding to your "own" (relatives, church members). This will be a fantastic opportunity to lead many fearful to salvation! Those who make the preparation-minded shift early and are able to supply others food will be heroes in the eyes of everyone they feed, plus they will be blessed disciples of Jesus: *"Blessed is that servant, whom his lord when he cometh shall find so doing"* (Luke 12:43).

God is desirous that we share new insights and wisely respond to His Word in simple faith; second-judging the Scripture leads to doubt and foolish inactivity. An Addendum added at the end of this book will present many ideas for preparing and creative thoughts for securing dwellings.

Two Human Reactions When Facing Disaster,: and How You Might Help Someone.

Studies[1] have shown that people react in one of two ways, when facing disaster. The first way is *Situation Awareness (SA)*, which brings about an alert diligent responsiveness; the second way is *Normalcy Bias (NB)*, which paralyzes a person's ability to respond. An illustration of these two types of reactions to danger was seen following a collision of two planes in 1977, above the runway at Tenerife, Canary Islands. Tragically, many occupants passed away on impact. A survivor however, told how he escaped as flames slowly engulfed the plane. Walking out, he passed by some healthy passengers, who might have stood up and escaped with him, but remained seated and quiet; they made no effort to leave the burning plane.

In the first moments of an emergency, either SA or NB begins driving a person's mental responses, and seldom changes thereafter. Normalcy Bias is a paralyzing mental state that many slip into during an unfamiliar dangerous or disaster setting. Estimates from research indicate that as many as 70% of people suffer NB when faced with a unique calamity. Since "everything was fine just a moment ago," NB victims take on a detached disposition, in which they do not react sensibly. Their thinking remains unconnected from their changed surroundings. Sufferers are unable to grasp clear meaning from what their five senses are saying. This causes them to misinterpret or ignore their environment, and historically all too often, ends with tragic results.

In contrast to NB, other people take decisive action with Situation Awareness. This has an opposite effect. SA is an energetic awareness during an emergency condition. The person operating in SA is rapidly evaluating each sight, sound, and smell in a dangerous setting. Up and moving, all his senses are on high alert; no dull-minded false hopes are at work. SA respondents are efficiently thinking, looking and speaking out their options, while on the move. Situation-aware people are committed to respond rightly; they often use their heightened awareness capacity, to direct others around them to the safe pathway. These are the ones who emerge heroes, with stories of great bravery, from catastrophic events.

Typically, little thought precedes the onset of the two mental responses, SA and NB. However, once activated, one mindset will rule the actions (or non-action) of a person throughout an immediate emergency circumstance. Likewise, a single mindset can govern decisions over a long-lasting distressed period. For example, the Nazi genocide of millions of Jews showed that most of a nation can be blinded with Normalcy Bias. For several years, Jews knew that their friends and family were being taken away, yet they refused to believe it, because in their experience, it had never happened to them. Stories are told, how that some living close to the tracks turned up their radios to drown out the wailing cattle cars of passing people. While they neglected reality, the same fate crept up and overtook them.

A bias is a tendency toward, or partiality for something. Sadly, when victims fall into muddled NB thinking, they often blindly continue doing their normal activities. Ignoring their dangerous surroundings, they seldom recover in time. Your responsiveness however, can help people in immediate peril. They can be jarred out of NB by shaking them or shout about the danger.

Longer term NB however, may be deeply seated in a person's mind, reinforced by the media and others in similar thinking. They will require patiently coaching a victim away from denial, into reality. Unfortunately, if friends, family, or the media has reinforced someone in an NB mindset, it might not be possible to convince them of danger, or that a response is needed. This certainly could be heartrending if someone you love is moving toward destruction. *Education and early preparation are the best preventatives for the confusion of short or long term NB.*

Militaries know the more practical training a soldier has, the better his responses will be in the stresses of battle. The airlines, as part of each preflight, hold up and unlatch a seat belt for all to see, followed with exit instructions. The purpose is simple, to provide fresh emergency awareness. Brief, less than twenty seconds, this short visual reminder is enough to trigger people entering peril to think: 'unbuckle', prompting them toward alert SA movement, and lessen the tendency of motionless NB.

The proven effective way for people to avert being overcome with Normalcy Bias, in a perilous cultural setting, is for them to receive previous instruction. Of course, more than early instruction is needed if faced with a long-term global disaster. Continued survival in the multi-year great tribulation will rely on a three-legged approach... being educated, making prior plans, and adequate preparations. Like showing the unlatching of the seat belt prompts correct immediate responsiveness, early knowledge is also vital for believers to plan situation-aware responses and make preparations for the lengthy dangerous times.

Sharing Is Caring

We can share what we know with others, even if the whole picture is not perfectly clear. From this massive new paradigm we will naturally have doubts and questions. Can you see now that sharing is a loving act of faith? Letting people count is God's tip-off, like unlatching of the seat-belt; you are acting wisely and being obedient to the Scripture: *"Here is Wisdom. Let him that hath understanding count."* To faithfully share this with people, you are not taking responsibility for them; you are telling them about a set of 2000 year old Biblical signs, that match with the prophecies. Emailing or posting pictures of this book's cover is enough for people to seek God for themselves.

Surely it will be more heartening to know your loved ones have read this simple informational outline now, in a time of peace, than to realize they are entering the chaos with no biblical clue of what is happening. Or just as bad, is the challenge if part of a family or church understands, while others are drinking down media lies, and either being led away or turning on each other. Imagine the distressing friction that could erupt, especially if long-term NB blindness and stubbornness become set, as we discussed! Certainly, this intense time would be an extremely difficult period to try to study and readjust major doctrine, or remain in a martyr's line, with paralyzing confusion about where God is in it all. Some will claim in the chaos: "I'm left behind, I'm not saved!"

"Who shall separate us from the love of Christ? Shall tribulation, or distress, or persecution, or famine, or nakedness, or peril, or sword? As it is written, For thy sake we are killed all the day long; we are accounted as sheep for the slaughter. Nay, in all these things we are more than conquerors through him that loved us. For I am persuaded, that neither death, nor life, nor angels, nor principalities, nor powers, nor things present, nor things to come, nor height, nor depth, nor any other creature, shall be able to separate us from the love of God, which is in Christ Jesus our Lord" (Romans 8:35-39). *"Cast thy burden upon the LORD, and he shall sustain thee: he shall never suffer the righteous to be moved. But thou, O God, shalt bring them down into the pit of destruction: bloody and deceitful men shall not live out half their days; but I will trust in thee"* (Psalm 55:22-23).

Trusting God in this setting, we should recognize that it may mean that we endure suffering along with perhaps, some form of unwarranted public shame and humiliation. Third-world-believers are all too familiar with persecution and public rejection. But abuse is a different worthy Christian honor than is typically preached in America. Perhaps, because the idea of persecution has not been relevant until now. Nevertheless, remember that our thinking must always align tightly with the Bible. The strength of our confidence rests in an unshakable hope in our Lord, without which, we might be tempted to doubt God.

John the Baptist lost hope. He saw miracles, but while facing death in prison, in despondency, he sent two of his disciples to Jesus to ask Him if He was truly the expected Messiah. The Lord in answer, performed many miracles that they witnessed, then Jesus sent them back to John with a strong admonition: *"Blessed is he, whosoever shall not be offended in me"* (Luke 7:23). We must not blame God for any difficulty we may find ourselves in.

Instead, may our outlook mirror Jesus's other disciples, who were also in prison, facing possible death, in Acts 5:40-42:

> "When they had called the apostles and beaten them, they commanded that they should not speak in the name of Jesus, and let them go. And they departed from the presence of the council, **rejoicing that they were counted worthy to suffer shame for His name**. And daily in the temple and in every house, they **ceased not** to teach and preach Jesus Christ."
>
> — ACTS 5:40-42

> "In all things **approving ourselves** as the ministers of God, in much patience, in afflictions, in necessities, in distresses, in stripes, in imprisonments, in tumults, in labours, in watchings, in fastings; by pureness, by knowledge, by longsuffering, by kindness, by the Holy Ghost, by love unfeigned, by the word of truth, by the power of God, by the armour of righteousness on the right hand and on the left, by honour and dishonour, by evil report

and good report: as deceivers, and yet true;
as unknown, and yet well known; as
dying, and, behold, we live; as chastened,
and not killed; as sorrowful, yet always
rejoicing; as poor, yet making many rich;
as having nothing, and yet possessing all
things"

— (II CORINTHIANS 6:4-10)

As we read those verses, we see that circumstances must NOT govern our outlook. The conflict potential is found in the first sentence: "***In all things approving ourselves as the ministers of God***" We must not lose our good perspective of our personal relationship with God... in other words, do not neglect your relationship with God. Circumstances are prophesied to turn against us, but God - *"who so loved the world,"* ***never will***... see Rom 8:31-39

6

SUMMARY AND PLEDGE

O ur loving Father has been forever a God of order.

Ignore temptations to be afraid, quote Bible verses and sing instead. Since the joy of the Lord is our strength, we must always be laboring to enter His rest, and maintaining a focus on His sacrifice of love. Relax, truly nothing is better than going to heaven! The things that will happen in the great tribulation are actually what has always gone on here on earth, since the days of Adam: people living and people dying. Only now it is in a merciful decision-demanding concentrated time period. As blood-bought Christians, ALL of our sins are forgiven. When Jesus died, He took the sins of the whole world away: Behold the Lamb of God, which taketh away the sin of the world (John 1:29). People need to hear the astounding news, that their life time of sins are not counted against them- if they only call on Jesus to save them!

Those of us who do end up in the ranks of waiting martyrs, must not listen to Satan's lie that it was because of our sin... Just the opposite; if we are standing there, it is God wanting to promote our encouragement and strength! It actually means God knew we could handle the special challenge. Resolutely following through is an opportunity to give confidence to those behind us, and finally, greater eternal rewards!

Except for the sign of Jerusalem being surrounded in Luke 21:20, it is not known if the pivotal "abomination of desolation" signal event will come abruptly with no prior warning, or if some publicly recognizable activity will enable us to discern, over time, that the momentous event is near, allowing us a little more time to notify others and prepare ourselves. The idea of it coming as a "snare" doesn't favor the latter (prior warning). How frustrating to have been warned early, only to face your wealth trapped in accounts, while you hold useless stacks of paper assets, have a garage full of toys, own a house full of finery with a meager pantry, and everyone you know in shocked surprise and equally unprepared. Wouldn't an ideal faith-response to this message be, to begin notifying others of these prophetic insights now, as you've learned them, while also steadily turning of all your non-necessities into food inventory, while food is cheap. Then you use the food to support as many people as possible, and finally end up (fully invested in others, aka- penniless) eating your last stored bites, as you look up to see Jesus enter your view on the horizon, or (blink) you are raptured? Because of God's Love for Us, This Will Be Possible!

Consider, Almighty God has shown us a specific detailed warning sign for the very first day that we are to start counting off the days remaining to endure; then 1334, 1333, 1332... what mercy! By calculating and collecting a food storage inventory, according to the 1335 days of the great tribulation, groups of disciplined

Christians working together could bring that food-planning ideal to reality! To determine the amount of inventory needed, count the number of people and the remaining number of days to endure. In other words, figure the total number of meals at one pound dried food/per adult/per day. Adjustments can be made, to either allow more people to enter and eat, or a method initiated for rationing remaining food. Having scales for weighing and managing food will be a must. God never ceases to care for his people even in unusual ways. The crafty Apostle Paul survived by being let down with a rope. Reconsider a weighty denominational obligation. Be assured the Lord is not applauding mindless religious repetition of any kind. He is only interested in a relationship with you, and you responding to His Word. Shedding dry monotonous religious traditions may facilitate your down-the-rope-and-outside-the-city type responsiveness. The prostitute Rahab saved Israeli spies with an outright lie that they had already left; it ended up saving her household!

Later the Apostle Paul noted in Hebrews 11:31 that she was saved by faith when "she received the spies with peace." Throughout history, we read of times that faith in God demanded bold action; the great tribulation era will be another such time.

Some Additional Thoughts, Observations and Cautions

Almighty God has allowed trials to prove or test people at all different points in history - often specifically with a lack of food and water.

Deuteronomy 8:2-3 tells of one such test:

> *"Thou shalt remember all the way which the LORD thy God led thee these forty years in the wilderness, to humble thee, and to prove thee, to know what was in thine heart, whether thou wouldest keep his commandments or no. And he humbled thee, and suffered thee to hunger, and fed thee with manna, which thou knewest not, neither did thy fathers know; that he might make thee know that man doth not live by bread only, but by every word that proceedeth out of the mouth of the Lord doth man live."*
>
> — DEUTERONOMY 8:2-3

For 2000 years, generations of people have lived in a comfortable, Grace "soaked" world. Do we dare now criticize God, if He has chosen to humble and prove what is in the hearts of earth's last living people, not for forty but a mere three and a half years or so? Will people calling themselves Christians deny Him and fall away, or is today's modern faith, in truth, an enduring faith that pleases God? It seems the Lord will similarly test us to see. We must never murmur against God.

There are other areas of concern that should be examined. Blind allegiance to a person or a denomination could ensnare ungrounded individuals and assemblies.

Our human tendency is to over-trust spiritual leaders. Don't do this. To grant people in authority, governing control of your humble vulnerable naturally submissive spiritual life, can leave you open to be easily misguided and manipulated. History shows that wayward spiritual leaders can mislead, take advantage of and herd people like sheep, and otherwise destroy followers, particularly when using the name of God. If what any man or group promotes goes against the plain reading of

Scripture, their ideas must be subjugated or disregarded. It is probably wise to reconsider your association with them. God's Holy Word is your and my first and final guide; it is both personal and reliable. How dangerous to rely on someone else to apply God's eternity-influencing meanings, blindly to your life.

The utmost extent to which we are to cling to God alone and if needed, go against even those in our own family, is shown in Deuteronomy 13:6–9. We will each stand and give account to God for our own words and actions, regardless of where our reasoning and motivation came from. On judgment day, "He said so," will sound pitifully shallow if it conflicts with, It is written. Eternal safety will always remain tethered securely to the whole counsel of the Scriptures."

Many people today are health conscious and pursue a quality of life that includes exercise and healthy eating. The Bible teaches that bodily exercise does profit a little. However, an over-focus on ones own well being in the face of danger, could easily switch to a preserve-myself-at-any-cost mentality. Hurting innocent people, to preserve your own life, is not justified in the Bible. In fact, the Bible teaches that we are to hate our life in this world, in comparison to our valuable eternal life.

We should be prepared to quietly give up our life if need be:

> *"Verily, verily I say unto you, Except a corn of wheat fall into the ground and die, it abideth alone: but if it die, it bringeth forth much fruit. He that loveth his life shall lose it; and he that hateth his life in this world shall keep it unto life eternal"*
>
> — (JOHN 12:24-25)

Our lives should be focused on the good of others:

> *"For to me to live is Christ, and to die is gain ... For I am in a strait betwixt two, having a desire to depart, and to be with Christ; which is far better: Nevertheless to abide in the flesh is more needful for you"*
>
> — (PHILIPPIANS 1:21, 23-24)

If death comes, it is far better gain, while self-preservation does not justify hurting innocents.

FAITH HAS FEET

Finally dear reader, for you to have this knowledge is like being the first to see a massive hurricane moving slowly toward your loved ones and acquaintances. They have not heard of this crisis coming. Please share this new information so others will not be shocked and confused should things begin earlier rather than later.

The Bible's description of the great tribulation is difficult to consider with any confidence of grasping the total impact it will be to the lives of those who face it.

Not being certain of a storm's exact direction or full intensity or precise timing, does not stop wise people from preparing their homes, when they know one is coming. Yet, once again, others can only do so if warning is given them by you, this won't be on any radio or TV.

Naturally, everyone will form their own response, or non-response; that cannot be your or my concern.

What if the recipients of your early notification had time to notify others, and in heaven - from your diligence, you find many thousands had time to consider, share and wisely respond? Certainly, when the tribulation begins, it will be much more invasive and disruptive than any weekend storm. Posting copies of this book cover can alert and inform.

Having this general knowledge, before the start of events, will allow individuals a time for reviewing their doctrines in light of a lot of new information. Most importantly, people can adjust and establish correct spiritual decisions. As well, groups can give attention to physical and emotional readiness and unity. Informing people can allow them to stay clear of the level of confusion and fear that those caught in the prophesied "snare," will most certainly experience. After notifying people, you could rest assured knowing those you care about, will have read (or for later - have it in paperback form) the clear warning God gives in Revelation, of the fiery eternal destiny of people who take the mark of the beast.

If you still do not get the purpose of this book, read Matthew Chapter 24 more closely. It culminates with verse 45 and 46, "Who then is a faithful and wise servant, whom his lord hath made ruler over his household, to give them meat in due season? Blessed is that servant, whom his lord when he cometh shall find so doing". We are to gather provisions for our households. Quick question, are you leaning at this point, toward an NB or SA approach?

The greater the number of people who see and understand these things, it stands to reason, the more of a fizzle the anti-Christ's fame and influence will be. Each of the prophecies will certainly be fulfilled, at some point, but how many more people can be prepared or at least not enter the tribulation utterly confused and at risk of being deceived? Can we as a body of individual believers show off to God, our unwavering zealous faith? Jesus posed an interesting question for our generation to answer: "When the Son of man cometh, shall he find faith on the earth?" (Luke 18:8). Sharing is not just thoughtful it is crucial for people to know about these needs early and to know how the Lord expects them to prepare to meet them. Emailing pictures of this book's covers can provoke interest enough for truth seekers to investigate.

The "Lord's Prayer" will certainly take on new and greater personal passion during this time: "Thy kingdom come. Thy will be done on earth, as it is in heaven. Give us this day our daily bread" "and lead us not into temptation, but deliver us from evil" (Matthew 6:10-13). Please pass out this information. The Bible is practical. God has pre-mapped these times for us. As His children, we love light and naturally despise living in the dark! The Apostle Paul writes, "But ye, brethren, are not in darkness, that that day should overtake you as a thief. Ye are all the children of light, and the children of the day: we are not of the night, nor of darkness" (1 Thessalonians 5:4-5).

Matching Evidence Warrants Careful Consideration

Some people, who began believing Jesus was the Messiah, mocked in disgust the resistant religious leaders who were standing nearby, challenging them:

> *"When Christ cometh, will he do more mira-*
> *cles than these which this man hath done?"*

> — (JOHN 7:31)

This is a pressing thought for us. How many prophetic signals should it take to convince us? If statistics were used to analyze the probability of these many diverse indications having occurred or that are now in plain alignment with Scripture to be fulfilled, the result would be considerably against the assembled information in this book, being a massive global coincidence.

Further, we understand that our loving heavenly Father is in control of this world; is it in His heart that He would allow such a broad array of odd identifying features to harmonize into a distracting deception, when His instruction to us repeatedly, is to be watchful for these very signs? Of course not. It might be uncomfortable but we must believe.

The Lord has prepared the Bible for all people to read, understand and obey. In the past, He has used fishermen, common tax collectors, and a general cross section of a culture to emphasize this. God has not subjugated the world to sit outside the doors of theological scholars, breath-holding for them to interpret the Bible and share with us. The real question, like all past transitioning people, will we dare to boldly believe the clear wording and indications given by God in the Bible, with child-like faith?

This book is written in sincere and simple obedience to the Lord's Revelation 13:18 command:

> *"Here is wisdom. Let him that hath understanding count... the number of a man."*
> *You have effectively fulfilled this prophecy when you counted his number. So what now? It is written: "and they shall place the abomination that maketh desolate. And such as do wickedly against the covenant shall be corrupt by flatteries:*
> *but the people that do know their God shall be strong and do exploits. And they that understand among the people shall instruct many"*

— (DANIEL 11:31-33).

Should we ignore this instruction, expecting a different prince later, one whose name calculates to 666, confirmed in two languages by the very calculating system common with John of Revelation? Are we to expect this other prince to have a lineage chart showing that he is a descendant of King David? Are we to imagine another prince coming, with secretive control of the majority of this world's food distribution, while exactly matching each of the many specific and odd identifying symbols of the Bible? Will another Jewish group arise, who eagerly anticipates a "messiah" son of David? Are we to expect this other group, having as their primary mission to sacrifice on an altar of stone, and to possess such a unique stone - made without tools, while also having as their primary goal to put it on the single spot on earth that God has chosen to perpetually put His eyes, His heart, and His name?

Should we expect this alternate group to replace the current people who have already proven their intent, by attempting to place the stone to begin sacrificing?

Having this knowledge, what excuses will we give to the Lord, if the abomination occurs and anti-Christ begins to reign, but we abandoned his Revelation directive to allow others to count, to recognize, and be ready?

An objective court would likely view the quantity and quality of these very unique prophetic fulfilled features, as evidence beyond a reasonable doubt. I'm watching, preparing, and sharing. I've determined that, as for me and my house, like Joshua, we will serve the Lord (Joshua 24:15). The very definition and nature of "biblical faith" is action that stems from belief in the words of the Bible. Moreover, faith is actually true faith only when it results in good works; remember that "faith without works is dead" (James 2:26). Now, its your move.

I have stretched my faith and written Dear Christian, Do NOT Fall Away, with no visible evidence of the tribulation starting. I've written during a time in my life that is peaceful, financially stable, and filled with the love of my family and a great church.

Please share the link to the book. Not everyone will purchase the book; that's okay. At least having the general information, people will have some understanding that can provide direction. If the tribulation "snare" becomes commonly expected, even if it begins years away, more of the tested people who are caught in it, can make godly decisions, prepare, and educate others. Knowledge of these things automatically makes you a leader!

*"My People Are Destroyed for Lack of
Knowledge"*

— (HOSEA 4:6)

An encouraging biblical directive is for us to actively watch for the Lord's return. Jesus speaks of His return in Luke 21:27 and 28: "Then shall they see the Son of man coming in a cloud with power and great glory. And when these things begin to come to pass, then look up, and lift up your heads; for your redemption draweth nigh." Since we are told to look up as events begin, and then watch as our "redemption draweth nigh," it seems that we will be able to actually observe over some period of time the coming of our redeeming Lord toward us! Again:

*"Behold, he cometh with clouds; and every eye
shall see him"*

— (REVELATION 1:7)

His appearing will begin "after the tribulation of those days... then shall appear the sign of the Son of man in heaven: and then shall all the tribes of the earth mourn, and they shall see the Son of man coming in the clouds of heaven with power and great glory" (Matthew 24:29-30).

What an overflow of contrasting emotions the great tribulation will have; the unbounded exhilaration of watching the physical return of our Lord Jesus Christ coming in the sky in like manner as He went, even while we endure rapidly collapsing surroundings. Jesus ends His sermon in Luke 21, regarding His Second Coming, with the command: "Watch ye therefore, and pray always, that ye may be accounted worthy to escape all these things that shall come to pass, and to stand before the Son of man" (vs. 36). Remember, a single cup of cold water given in the Lord's name receives a reward.

Giving this information to people you may not even know, could answer Jesus's question,

> *"when the Son of man cometh, shall he find*
> *faith on the earth?"*
>
> — (LUKE 18:8)

YES! After 2000 years, the Lord has multitudes of people around the world, who both claim they believe the Bible, and have the independent faith to recognize prophetic signs and respond.

Again, it is written:

> *"They that understand among the people shall*
> *instruct many"*

> — (DANIEL 11:33)

This information is so important; yet it's hardly being addressed or even considered among believers. It's too volatile for TV, too seismic for radio. Yes, it can seem overwhelming, and yes it can take your breath away.

God lives inside you and me, speaks through us, and acts through us if we are a born again member of his family. Dear Christian, stand up straight: you and I are children of the King of kings; we must all live and talk like it! God has given the world signs, like counting the number of anti-Christ's name, to provide an introduction and verification to end-time understanding. By these, He intends the whole planet to be alerted! Few people recognize when true earth-shaking moments occur in their lives. Can you see this moment, and will you participate in an informative effort?

The depth of despair that our clueless and confused relatives and the people of the world will feel, if they enter the great tribulation ignorant, would be beyond comprehension! Such confusion could lead to ensnarement by a satanic, deceptive system. It takes faith when things are peaceful to make preparations and distribute this information. Please do not let doubt or fear prevent you from boldly helping as many people as possible, to at least be aware of these things.

The tribulation will come as a snare upon us all. God has been and is forever merciful:

> *"Except that the Lord had shortened those days, no flesh should be saved: but for the elect's sake, whom he hath chosen, he hath shortened the days"*

> — (MARK 13:20)

As This Book Is Shared, People Will Have The Opportunity To:

See the Lord's love in history, the warning signs, prior to judgment!

Know that someone cares for them and is praying for their endurance!

Trash conflicting beliefs of an ultra-ancient aged earth, mythical monkey men, and an incoherent bumbling Creator!

Embrace an independent faith, cleansed from religious ritual and scientific deception!

See photo evidence the Lord spoke accurately to our enslaved forefathers to free them, and trust that He is speaking truth to us now!

Gain confidence in the words of the Bible from the very first verses, and pray for his or her simple salvation!

Clarify, Biblical fact from fiction, when the rapture may occur!

Accept and Value the unlimited power of God, to lead them daily!

Trust their Creator's love even in chaos, to lead us to His eternal home!

Obey God's compassionate prophetic command to "let" others with "understanding count the number of a man!"

Share this knowledge with others, and rest, knowing friends and family will not be clueless in catastrophe and calamity!

Make crucial decisions wisely from a grounded faith, rooted in a comprehensive worldview!

Know to estimate, gather, and give their households portions of meat, to endure the fixed written tribulation time period!

Have an opportunity to study and apply this detailed Addendum of ideas and preparation insights!

Face chaos into martyrdom if need be, knowing all creation, particularly man's prophesied calamitous closure, validates their Biblical faith!

Every generation tends to believe that they will not fail God, as have those previously; it is our human nature. Pride and hubris however, take no one far with God.

Proverbs 16:18, Faith, Hope, and Love make up our spiritual personality now and in the future. Only in heaven will we know the final help this information was to others. God is our Father, and Jesus is our Lord and Savior, we have the Holy Spirit living inside of us – the same power that raised Christ from the dead! You and I will remain forever in God's family, among God's chosen! Rejoice! Heaven is now close at hand!

Believer's Final Earthly Pledge

Whereas I, being a believer in the Lord Jesus Christ, do separate away all that is in my heart on this earth that I have counted valuable. I do turn now and embrace with all my affection and obedience, the Lord Jesus Christ and His Word.

And Whereas, in doing so, I relax and release all things freely without looking back. I pledge to God, to my family and friends, from this moment on, to walk forward daily into the loving arms of God, looking without fear for His fulfilling conclusion to my life.

Therefore, from now to my life's sacrificial or natural end, no human, place, thing, or doctrine, will I allow to turn my heart against loyalty to my salvation, bought by Jesus with his own precious blood on a torturous cross. I fully trust that by God's Grace, and Jesus' intense payment:

1. *I will NEVER take the mark of the beast.*
2. *I will Hold my focus exclusively on the joy awaiting me in heaven, and,*
3. *I will Finally be resurrected by God Himself, into His eternal glory.*

In Conclusion, I boldly vow, that by God's grace, I will hold fast to faith in my Lord and Savior Jesus Christ, endure the great tribulation, and be a loving encouragement to those around me on the journey.

_____ *Believer's Name*

_____ *Witness*

_____ *Witness*

ADDENDUM: PREPARATION IDEAS, YES YOU CAN!

Unwavering Christians Will Shed Worldly Habits and Desires, to Endure.

It seems clear that we are likely the Christians scheduled to live through the great tribulation. If true, we are prophesied to undergo hardships. Faith drawn from personal time alone with God and His Word is key to successfully endure. Sin, abrasive attitudes, and bad habits cause confusion and inhibit wholehearted communication with God. More courtesy, kindness, and deference are needed now than ever! It is written: *"only by pride cometh contention"* (Proverbs 13:10).

We know God resists prideful people. Fasting and reading His Word can help transform abrasive self-centeredness by erasing doubts, controlling emotions, redirecting destructive thoughts, and especially stifling fear. Doing so exercises spiritual control over the body, the thought life and the tongue. This really works and can help reform thought patterns and set up personal thought boundaries. It is very important to practice examining and adjusting, what you are thinking about moment by moment.

You generally move toward and receive more of what you dwell on. The longer a fast is, normally the more retraining progress is made. Fasting can be integrated as part of a routine during the week; combined with daily Bible study and prayer, any Christian can enter a highway to maturity, especially in the face of oppressive surroundings. Fasting simply reduces fleshy mental activity and quiets gastric influence - to enable Holy Spirit to control and offer calmness. This will allow better communication to and from God, but has nothing to do with salvation.

Remember, the apostles Paul and Silas, who when shackled in prison and facing a trial for their lives, took action to maintain joy; their response was to vigorously sing. Idle lips and hands are the devil's playground, so it is important to stay productive. This is no time to play games with God. It has been said, God has no favorites, just intimates. Be sure to stay close to Him daily, first thing!

As Christians, our mental and emotional stability is not dependent on our creature comforts or maintaining hygiene habits. With food and clothing, we should, according to Scripture, remain content. In one sense, we might consider this time, a draft by our Lord into a short stint of mission-field work. Remember, that a majority of the world has always lived braving grim poverty. We have lived bountifully blessed, though we were never better than the humble African carrying the water pot on her head for miles along a dirt road. Among this world's challenged people, we find our perfect example, that of the bold, mature, sparsely supplied Christian missionaries, living meagerly in obscurity to simply serve others and win souls to Christ!

It will be a selfless time, in which we must be humble, *or we will be humbled by the mature among us.* Maintaining hope is important, whatever our circumstances. Our emotional well being will only be as healthy as our current state of faith; *remember this.* Faith and fear function the same; one grows as we believe God, the other from believing the enemy. Monitor what you are cultivating in your thoughts.

The strength of our faith rests on hearing and quoting the Scriptures often, while fear feeds on negative reports and vocalizing our worst thoughts. *"Faith cometh by hearing and hearing by the word of God"* (Romans 10:17). *"They overcame him by the blood of the Lamb, and by the word of their testimony; and they loved not their lives unto the death"* (Revelation 12:11). As important as physical food, during the coming trials, is the need for a consistent intake of heavenly bread, and taking time to quietly be with our Creator. It is written: *"The effectual fervent prayer of a righteous man availeth much"* (James 5:16).

The Effective Christian Leader, Is the Biggest Servant!

Any leader over your family should be selected carefully. Are their ideas rooted in a desire to maintain control, money, pet doctrines, or a position? Do they believe more in what *they* say, than what God's Word says? Do their opinions and decisions originate from traditions, emotions, preconceived ideas, fiction, or superficial 'half-verse' doctrines? Or instead are they wholeheartedly directed by the whole Bible only? Do they compare people and evidence, to what is described will be the end time indications? Does scripture move them to make changes, and when they realize they should do they follow through? Would they be agile and responsive enough and willing to turn and break with all current plans, life routines, and redirect monies and time, to lead on in survival? Be aware, the callous-handed quiet plumber might lead more practically than any others in this new harsh season.

Since God resists the proud, are they humble? Good spiritual leaders during Grace may or may not be good leaders during Judgment. This position of influence in your life could have eternal impact on you and your family. Don't let anyone split family allegiances. One telltale sign of unstable pride in a person, might be if when new information is presented to them, they reject it without diligent, prayerful consideration; or do so without sound biblical reasoning. Don't risk your life under selfish or prideful leadership. Be sure leaders over you and your family earnestly have your best interests at heart, and also that they are qualified in the new environment to lead. Wives be wise, your man has only faced bugs on the wall, this is a voracious toothy dinosaur-like era. He needs respectful confidence-building encouragement, to be his best.

Frankly, no one would be wise to rely on someone else, blindly, to store food for his family. Just as there are stories of lottery winners behaving uncharacteristically, in the opposite spectrum of sudden grim survival, dependable people may change unpredictably. Clear communication is vital, as even simple misunderstand-ings could cost lives in some cases.

Make all group agreements in writing. List names, detailed ideas, locations, spending, and portioning; if signed by all involved, this will firm up and clarify mutual understandings. This will also facilitate enforcement, including a decision to receive someone not listed - if their inclusion would breach food availability.

Practical Ideas on Buying, Storing and using the Essentials:

Collecting and storing food as did Joseph in Egypt, is biblical, but it takes lively faith during times of plenty. Hoarding food to sell later at high prices however, is cruel and not biblical. Storing too much is probably not possible unless transportation challenges arise, while storing too little food or including too many people, could lead to intense stress. Perhaps the Lord allowed Y2K as a trial run. Despite those concerns never manifesting, all of God's prophecies about the future are absolutely certain at some point. They are tied decisively to His indicator verses.

Today, many people have promoted gold and silver hoarding. However, buying cheap food early, with available cash, is a wiser choice. Ezekiel 7:19 speaks of the time when silver and gold will be tossed away as useless. Food will be king; load ten carts of food and hand your checkout clerk a bag of silver coins or a one ounce gold Eagle. Even if their values have tripled, the clerk does not know their current worth and cannot process the order.

Who's In and Who's Out, and How Would You Know?

The hardest step in any journey is the very first. When you have considered your direction, go ahead and make at least a small baby step, even if as a token or mental ascent, to show respect of the Bible verses. Prioritizing your life to include preparing need not be an emotionally stressed decision. It actually makes good sense anytime. How hard is it to start today, by simply shelving extra cans of tuna, to kick off your pattern of prayer and preparation? It is said that the early church, to identify fellow believers, would draw an arc in the sand when first meeting a supposed Christian. The other, if a believer, would then draw a lower arc from the front, down opposite and crossing up the back of their arc, to form a simple fish or ichthus, a symbol of Jesus's multiplication of the loaves and fishes. We can make our up-to-date identity confirmed with the helpful generic question, 'Got Tuna'?

Only in the last hundred years, with the developing business practice of just-in-time inventory management and with grocery stores and restaurants ever more convenient, have people not considered long-term storage and rotation of food a useful habit and an important type of savings. Even so, while all of our choices must carefully align with scripture, the Bible provides little in the way of direct instructions for how the Body of Christ, an individual Christian family or church, is to endure the great tribulation. Steps 1-6 are just not laid out for us. We must study the Scriptures-believe them, glean ideas and principles, and exercise our faith. This book could not be exhaustive in this regard, but is meant to offer a starting point for information, dialog, investigation and action.

As the coming worldwide horses of Revelation chapter 6 approach, our trust must be in God alone. Remember, no Christian will be able to buy or sell in the commerce of that coming day, because no faithful Christian will be taking the mark of the beast. If you get this information late and have already taken the mark of the beast, you must somehow remove it. After Peter denied the Lord three times, even with cursing, he was asked later by Jesus three times if he loved Him. Scripture speaks of some kind of worship when the mark is taken. What can you say to God, to revoke your denouncement of Jesus, when you took the mark? God knows what each persons best plans might be and when we are to start them. He is capable of showing us the details.

From here, stay in peace and prayer. Food prices are prophesied to go way up; buy early before the panicking begins. Buy locally to avoid delivery delays, which are bound to increase from weeks to months and then finally stop without the mark. Buy in bulk, the more, the cheaper. Grain elevators, local farmers, wholesale marketers, and warehouse stores sell at the cheapest rates (the latter will probably be the first to limit sales). Within 1-4 hours outside of most cities, you'll find grain elevators. Buy in cooperation with other families and within churches to get larger quantities. Consider a commodity futures purchase of 40,000 pounds of grain, through a commodity broker. Futures options purchases work like a down payment and will hold your price fixed for months. Most all foods can be stored for long periods, if preserved correctly.

Believers, with robust faith and available money, might, at the right moment, use grain commodities as their last worldly investments. Imagine earthly wealth changed directly into stockpiles of gift food, converting "filthy lucre" directly into eternal heavenly rewards: *"Blessed is that servant, whom his lord when he cometh shall find so doing"* (Matthew 24:46). To haul quantities cheaply, use open trailers for access to a grain chute. It may be wise to pay with cash.

The source of the very cheapest, field-grade inventory is at your local livestock food bagging facility. Our Lord Jesus thought it okay to eat field grade grains (see Matthew 12:1). They have light debris present, but the grain stored in these places is kept bug free in silos. The food is wholesome and edible, but has not gone through the sorting processes required for regulated human retail. Corn labeled for deer is nutritious and perfectly edible. Certainly $7 to $10 per 50 lb. bag of easily cleaned grain is more affordable than the triple cleaned $40 to $60, 50 lb. retail bags sold elsewhere. Sunflower seeds – as bird seed is cheap and can be sprouted in a couple of days, and eaten for the enzymes and roughage.

Be creative, soybeans, rice, corn, oats, and all food of every grade and in every form will quickly become extremely popular. *It will keep you and your loved ones out of the destructive lines* (the Bible speaks of beheading Christians in this era). Remember Noah, who while preparing, warned his neighbors. He looked like a fool to them, but later had his preparations and provisions cried out for, by all outside the ark. God closed the door of the ark Himself, probably to prevent Noah from opening to people he knew, calling his name, crying for his provisions.

The wisely prepared wedding virgins (your group) will have adequate supplies. The parable instructs us that our sharing is to be likewise strictly limited, or not at all. It's easy to imagine the extreme challenges people will face, who have *allowed their food inventories to become known.* Surrounded with dire need, limiting the giving Christian hearts within a group could be a source of conflict. Perhaps those who insist on giving food away against counsel should be required to start by giving their next week of meal portions.

I hope you are seeing how important the need for sharing this information. It would be so sad to tell people who we know, but who never heard this information, to *'go and buy for yourselves'* as that parable depicts. One more thought on sharing: it might be easier to share this information immediately while you have no idea what preparations *you* will make, rather than be busy in work with a plan and face the need to deflect inclusion requests as you prepare. 'I don't know what we'll do,' is said honestly.

Apportioning Limited Funds

By beginning with lower cheaper grade foods, funds-limited people can rapidly build a survival inventory. As time and money allow, the more pleasant foods, spices, and condiments may be added. Dehydrated foods are purchased or made yourself. Sprinkle frozen fruits or vegetables over cookie sheets, opening the oven 1" and cooking at 145 degrees F overnight. It is less expensive than canning, less work, and takes up a fraction of the storage space. A bushel of tomatoes for example, equals 32 quarts if canned. Dehydrated, ground and stored, they will take up only 2 quarts. No nutritional value is lost storing dehydrated food. All meats are salted and dehydrated into jerky in a similar way and stored effectively as well.

Start now copying and filing internet ideas along with wholesale supply addresses and usable recipes; be ready should power go out. Buy and butcher cattle or any livestock to dehydrate the meat for a cheap means of building inventory. Yeast is made with one tbsp. honey (or sugar), one cup water, and one cup flour. Mix well and set open in a warm area. It will attract yeast spores in the air and will ferment into leavening. Pasta is cheap, provides needed protein and carbohydrates, and keeps for two years with no preserving. Eggs are also cheap and a great source of protein. They are basic to many recipes. Nightly, beat a dozen or so and place them on a fruit-leather tray (or cookie sheet with parchment paper) at 145 degrees, crack the oven door overnight. In the morning, grind in a food processor with salt and store the eggs in powder form. Acquire unbreakable, easy-to-clean pots and stainless cookware and utensils for your anticipated group size.

Preservation and Storage

Different methods of preserving and storing foods can be used; ideas here again are by no means exhaustive. The simplest and least expensive considerations will be discussed. Grains and beans must be stored whole. Ground as flour, they do not hold nutritional value well, and their oils may turn rancid stored long-term. A hand operated grain grinder is a necessity for daily portions. Honey is usually bought and stored in five and sixty gallon containers, needs no preserving, and will last indefinitely. Sometimes honey does crystallize; if that happens, simply warm it to liquefy; this will not harm flavor or nutritional value. Salt is stored in sealed dry containers.

Whole grains are preserved inside new trash cans, buckets, or in bins with Mylar or Ziploc bags. To kill any existing bugs and any that might develop, sprinkle diatomaceous earth (from a pool supply) over grains and beans before sealing. About 1-1/4 cups for each 5 gallons of food is enough. Diatomaceous earth (d.e.) is cheap, effective, harmless to animals and humans, and need not be cleaned from food before eating; some people use it in smoothies. It actually provides minerals. Also for long-term storage, an inert gas, such as nitrogen or carbon dioxide from dry ice, will displace oxygen in the containers and is another effective way to prevent bugs from developing. Fill the buckets with moisture-free food and slowly fill with nitrogen over a five-minute period to prevent the gas from mixing with air. If you use dry ice, put a 4 x 4-inch x 1/2-inch thick brick on top of the food. Leave the lid on loosely for 12 hours and then carefully seal the container. The above gasses are heavier than air and fill from the bottom up. Tap the container as you fill to release air bubbles.

A concern during these times might be about keeping food hidden. The use of 3" or 4" diameter PVC pipe, for food storage, will allow burial of inventory. Cut and clean manageable lengths of new pipe; cap one end, fill with dry food and treat with d.e. and one of the above gasses, then glue closed. These can be buried side by side in above ground 'gardens' covered with mulch, stored in the attic, etc. Canned foods have a shelf life of about 12 months before the nutrition value drops to near useless, so eat these first. Acquire fishing nets (seines), hooks, and line if you have water within biking distance. It is conceivable that a family who has nothing but a fishing net, who lives next to a large body of water, could survive on fish alone. If you have no net, purchase spools of nylon cord and download instructions from the web to make fishing nets. Trot lines are simple and very effective as well. Study your locales hunting, fishing, and wild edible plants to have at least something to suggest to those you cannot welcome to your table. A helpful website dealing with long-term food storage and use is www.provident_living_today.com. Here, you can learn to gather, prepare, and store food for an adult for one year for about $200.

Water:

The simplest and cheapest water treatment method, aside from boiling, is to use household bleach. A small amount can supply anyone with quantities of safe drinking water. Be sure sodium hypochlorite is the only active ingredient in your bleach. Avoid soap additives and phosphates. For 1 quart of water, mix 2 drops of bleach; 1 gallon uses 8 drops; 5 gallons needs a half-teaspoon; and for 60 gallons, use 1 ounce. If the water is murky, as from a lake or other surface water, use twice as much bleach. Vigorously mix and allow to stand for half an hour, before using. Fabrics and coffee filters can clean out heavier debris.

Try to prepare each days water supply the day before. The bleach will evaporate out, leaving safe non-bacterial water. Hammering a special screened pipe into the ground and using a hand pitcher-pump makes a shallow 25 ft. well; check with your local hardware store for supplies needed. To store water, consider putting an above ground pool in a garage, covered with a tarp. Store a minimum of one gallon per person, per day, with half for drinking and half for washing. Sometimes water districts allow one free pool filling per year, check with them.

Energy:

Without ability to buy electricity and gas without the mark of the beast, Christians may be hard pressed to cook and heat an environment without prior preparation. A number of alternate sources of energy are available. An old wood or coal burning pot-belly stove can be vented up a fireplace chimney or out a window, for heating and cooking. Your stockpile of wood, coal, or oil kept inside will not be vulnerable to thieves or act as a display to others that you are well prepared. Sterno cans are great as a backup for cooking. In a pinch, 1-2 hours of 'canned heat' can be made using a folded sheet of newspaper pressed into a tuna can. Pour paraffin over the paper wick and light.

A stored measure of cheap charcoal briquettes and a small grill also makes for handy cooking. Another ideal fuel, for more comfortable daily use, is propane. It comes in 20- and 250-pound tanks and also in 500-gallon tanks. The gas can run a propane range, propane lanterns, and a small gas generator fitted with an appropriate carburetor. New propane heaters are now available that operate 100% efficient, and need no ventilation. Also, search Biogas Digester. The Internet has reams of information to construct a waste-based methane energy source for heat, lights, and cooking.

In severe cold, nothing can take the place of quality bedding and well-fitting, layered clothes. For lights, kerosene is readily available, but does not store well beyond a year; use 100% paraffin-based lamps instead. Consider buying hand-crank flashlights and radios. Tea light candles are a good idea as a backup, but hardly work outside. A large quantity of batteries and flash-light bulbs will be useful.

Be the first-in-fashion on your block, with new tribula-tion china. Select easy to clean metal camping bowls and cups. Also have plenty of lighters, or a scrape-lighter of magnesium and steel. A stove top pressure cooker is a most efficient use of cooking energy. Collect tools, tapes, fasteners, lumber, wire, glues, fabrics, and all sewing items. Everyone must become a handy-man/woman and a conservationist in this temporary retro-pioneer culture.

Medical, Hygiene, and Miscellaneous Concerns:

Basic hygiene could be a luxury in the tribulation days. Have enough soaps and disinfectants available. We Americans can live on far less than we are accustomed to. Toilet paper may seem like a necessity, but if funding is tight, boxes of cut up old clothing from Goodwill will do fine. Certainly everyone's personal standards will be *unseated* in many ways from a lack of utilities. Without a sewer system to carry away waste, old-fashioned night pots could substitute. In good times these ideas may seem like silly advances. Dedicate a 5- gallon bucket and lid for a potty. A toilet seat may be placed directly on the bucket. Replace the lid after each use to contain odors until the contents can be buried. A good first-aid-kit can be a lifesaver; store triple antibiotic ointment, hydrogen peroxide, and bandages. Super-glue is useful in place of stitches. A book on backwoods or military emergency first-aid procedures would be very valuable if needed. An hour of stress can deplete the vitamin C in the blood; ascorbic acid preferably with rose hips or acerola (C powder), will bolster the immune system. Prepare for specific needs; gather feminine napkins and personal medicines. If a baby is present or expected, have plenty of rash cream and fabric diapers. Patience becomes an extra challenge with an unhappy baby. Homemade laundry soap is easy to assemble: heat 6 cups water, add 1/3 bar of FELS laundry soap-grated, boil 15 minutes, then remove from heat. In a separate three gallon bucket, add 1 quart hot water, a half-cup Washing Soda, and a half-cup Borax; mix all, fill to the top with cold water to make 2 gallons of mixture; the soap will gel. Use a half-cup per load; the cost is pennies per gallon.

Gun ownership and use is of course, a personal choice. At the end of Jesus's ministry on earth, he told His disciples to sell garments and each buy a sword (Luke 22:36). Our Lord never promoted aggression; He was not likely introducing a new macho fad, a military fashion statement, new policing duties, or providing the disciples with body-building weights. So it is reasonable to think His command was to allow them for self-defense. Soon after, when Peter cut off a man's ear, he was scolded for his bad timing not lectured on passivity (see John 18:10-11). In Psalm 144:1, David's hands were taught to fight. I will not stand by helplessly, while my belongings are robbed, or my family is accosted. God put the sense (and teeth) in even base animals, to protect their young. Something as simple as pepper spray can allow a quick and harmless getaway from an angry beggar. The idea of stockpiling huge amounts of ammo, for a primary defense, shows much courage, but is probably shortsighted, if marauders arrive. Becoming hidden is a much better idea. Prepare your dwelling. Pre-cut plywood nailed or glued over windows and doors notifies intruders: this location is fortified and will be hard to pillage. Painted warning signs make known: stealing from here will be resisted. Fences of any kind provide a barrier. As persecution increases, men who are not shaven are obviously not buying or selling and do not have the mark. A supply of razors may help avoid unwanted attention. Christian men have a God-given mandate, as well as an instinctive drive to defend their family. The best protection may simply be in quietly hiding your family well.

Being Thoughtful and Creative with All Options Available

Everyone will likely have the same timing to their reactions; turn stocks into cash, buy bulk food, acquire propane tanks, secure dwellings, etc. The best response will be to stay ahead of the stampeding masses. Wicked governments rule solely to maintain their all important population control; justice is irrelevant, and in fact, injustice can be an evil tool to root out and eliminate non-cooperative people. So, don't focus on the evil selfishness of leadership – it's a waste of your godly mental energy, and loud resistance will make you a target. News will be overwhelming daily...I recommend Christians avoid watching news, at least in depth. Rest assured their wicked deeds will meet God's full wrath. Instilling fear and forcing compliance is their game plan. Not to start the new doctrine that every lie is okay, but Rahab in the Bible, lied to her wicked city police, to secure God's spies; the Lord rewarded her and saved her family from destruction.

This current global regime will be wickedly pushing a mandatory mark that works against God's authority, to destroy billions of people- eternally. Civil disobedience is not always wrong. It can actually become an obligation for Christians to deflect, sneak, hide, conceal, lie, disobey, fight, undermine, subvert, etc. Otherwise, living in compliant obedience would be honoring, leading others, and thereby validating, the satanic leadership and vision - by conforming. Those who do comply are contributing support to the anti-Christ's destructive influence. Primarily, we must work quietly around any measures that punish stocking up food, by being guarded while we go forward gathering long-term provisions. The blessed servant, remember, will have food and be providing portions for his household, when Jesus comes back. Foolish people will not.

The verses in Joel and Revelation, of the 200 million man army and the cross-continental burning the Bible says they will do, appears to be a major aspect at the end of the tribulation period (Joel 2:1-11 and Revelation 9:12-19). It is written: *"Immediately after the tribulation of those days shall the sun be darkened, and the moon shall not give her light, and the stars shall fall from heaven, and the powers of the heavens shall be shaken: and then shall appear the sign of the Son of man in heaven: and then shall all the tribes of the earth mourn, and they shall see the Son of man coming in the clouds of heaven with power and great glory. And He shall send his angels with a great sound of a trumpet, and they shall gather together his elect from the four winds, from one end of heaven to the other."* (Matthew 24:29-31).

The Bible further declares, *"When the wicked rise, men hide themselves* (Proverbs 28:28). As your neighborhood becomes dangerous (see Revelation 6:15-17 and 9:15-18), *"A prudent man foreseeth the evil, and hideth himself; but the simple pass on, and are punished"* (Proverbs 27:12). So, the wise hide and the simple suffer for not hiding.

With those warnings in mind, notice in Isaiah 26, the Lord goes all the way, with compassionate forewarning and distinct directions for people to *"hide thyself as it were for a little moment and to enter thou into thy chambers until His punishment of the inhabitants of the [whole] earth is overpast"* (Isaiah 26:20-21). Look at how our Lord again, mercifully speaks to *"my people"* what-to-do directions! It can't be any clearer than this. The verse starts off with strength-giving encouragement about the joyous end-of-the-world resurrection of earth's righteous dead: *"Thy dead men shall live, together with my dead body* (Isaiah's dead body) *shall they arise. Awake and sing, ye that dwell in dust: for thy dew is as the dew of herbs, and the earth shall cast out the dead. Come, my people, enter thou into thy chambers, and shut thy doors about thee: hide thyself as it were for a little moment, until the indignation be overpast. For, behold, the Lord cometh out of his place to punish the inhabitants of the earth for their iniquity: the earth also shall disclose her blood, and shall no more cover her slain (Isaiah 26:19-21).*

The resurrection includes Isaiah himself; he will be raised after *"the indignation be overpast."* Thus, he shows that the timing for God's people to hide is intended to be during the final tribulation era, just before the final all-inclusive resurrection of all humanity's dead. Isaiah and other Old Testament saints have not been raised yet. We are to hide *"until the indignation be overpast."*

God will *"punish the inhabitants of the [whole] earth for their iniquity"* and for all the slain undisclosed innocent blood that has been shed on earth. The punishment will take some time to accomplish, while we are forewarned to *"hide thyself."* Naturally everything above ground is easily seen, searched and pillaged, so, we will need a good hiding place. We don't have caves in most parts of the world. Digging a chamber, or as the Bible describes elsewhere, a "*den*" (Hebrews 11:38 and Revelation 6:15) under the slab of a home, an empty building, church, etc., could be used to remain hidden and for emergency living. Hiding underground worked for Saddam Hussein to live and elude detection for many months, even as our army's search '*overpast*' him.

A Particular Prophesy in Isaiah 54, Provides Pledges of Our Lord's Reassuring Kindness Toward Us, His Bride, During The Closing Tribulation "moment"

Pay special attention to the tender tone of this message from God to His bride, meaning us - His covenant partners of peace. The context is the final tribulation "*moment*." Below, End-of-the-World emphasis is **emboldened**:

"For the Lord hath called thee as a woman forsaken and grieved in spirit, and a wife of youth [Lamb's wife: bride, see Revelation 21:9], *when thou wast refused, saith thy God. For a small moment have I forsaken thee; but with great mercies will I gather thee* [resurrection]. *In a little wrath I hid my face from thee for a moment; but with everlasting kindness will I have mercy on thee, saith the LORD thy Redeemer.* **For this is as the waters of Noah unto me; for as I have sworn that the waters of Noah should no more go over the earth;** *so have I sworn that I would not be*

wroth with thee, nor rebuke thee [see Matthew 24:37-38]. **For the mountains shall depart, and the hills be removed** [see Revelation 6:14]; *but my kindness shall not depart from thee,* **neither shall the covenant of my peace** [our eternal salvation covenant with Jesus, our Prince of Peace] **be removed, saith the LORD that hath mercy on thee.** *O thou afflicted, tossed with tempest, and not comforted, behold, I will lay thy stones with fair colors, and lay thy foundations with sapphires* [see Revelation 21:19]. *And I will make thy windows of agates, and thy gates of carbuncles, and all thy borders of pleasant stones. And all thy children shall be taught of the LORD; and great shall be the peace of thy children"* (Isaiah 54:6-13).

Perhaps an analogy can shed light on the brutal contrasts in this passage, and help clarify the loving heart of God toward His *"tossed with tempest, and not comforted"* wife. Imagine intruders have broken into a family's home. The protective husband (God) scuttles His frightened bride and children away from Himself, so that He can turn and do battle (*in a little wrath I hid my face from thee for a moment*). He warns them they will be left anxious and not comforted for a time, but reassures His bride that His covenant with her is always secure and will *"neither... be removed."* He tells them that He will make the clash be over quickly *"in a small moment"* and that He will be back to rescue them! As He departs, He encourages them with special lavish promises, to build them (you and me!) an extravagant new home, *"lay thy foundations with sapphires,"* where *"thy children"* will be well taught and peacefully cared for. The reference to caring for our children is especially thoughtful as verses elsewhere, speak of grief for the nursing mothers, in the great tribulation.

My analogy of course, falls far short of the many promises throughout the Bible made to Christians who enter heaven's glory. Nevertheless, in this passage our loving Lord prophetically comforts, encourages, and strengthens His bride, who will feel neglected, tossed and in tempest. She will face the final tumultuous times and endure until, *"with great mercies will I gather thee."* Our Lord provides bountiful promises for us and our families. He assures us that afterwards we will have a marvelous peaceful lavish eternal heavenly future. Again, we see here that God is aggressively evoking hope in us that we might endure, for the joy of heaven...

Dear fellow believer, we have nothing with which we can compare the great tribulation environment or circumstances. Neither can we know to what extent our corner of the world will be affected. We must go forward with biblical principles for our responses and preparation ideas. *"A prudent man foreseeth the evil, and hideth himself; but the simple pass on, and are punished"* (Proverbs 22:3).

Biblical depictions were the source for these bold ideas and considerations. Perhaps older retired people with money for food, coupled with younger folks, who have strength for digging, make good groupings. The extracted dirt formed into crop rows serves triple duty, by providing hidden living chambers while concealing dirt extractions, and producing needed food.

The nursing (retirement) homes are filled with moneyed people who will not want to take the mark. Go in and make friends; give out copies of this book, and return in a week to discuss with any who would join and participate. You could be an answer to their prayers!

When events begin to happen, it will quickly become more difficult to prepare. Pray, pray, pray. The preparations you think you would like to have accomplished, begin to at least plan right away, even in small ways. Openly discuss all ideas and unreservedly share thoughts. "Hey, I read this book that documents really rough times are likely close at hand; here is a copy. If you agree, I would like to work with you to keep our families safe." *A man's heart deviseth his way: but the Lord directeth his steps"* (Proverbs 16:9).

We make our plans, and God promises to direct our steps. Notice God does not direct our plans, though we are to make them. But He has committed to personally direct our steps made from our plans. Immobile objects can't be steered and they make no progress. Now more than ever, the head of every house must remain at his post, hear from God, and be decisive. Frankly, it takes little imagination to consider what a hoard of armed soldiers do when they sweep through a civilian population.

A Brief Thought for Wealthy Christians and Leaders.

What an opportunity this could be for Christian leaders who have promoted the pre-tribulation rapture doctrine, to 'redeem' themselves within their congregation, using one of the biblical applications of God's tithes: *"Bring ye all the tithes into the storehouse, that there may be MEAT in mine house and prove me now herewith, saith the LORD of hosts"* (Malachi 3:10). God has always fed His lead servants with His tithes. As God directs, don't shrink at your flocks upcoming need! Imagine the response to a pre-tribulation rapture promoting pastor (or any pastor, leader, or wealthy person) who enters his hungry congregation, after the tribulation is recognized, and announces he has X-thousand pounds of food to divide among his people! Can you help prepare for your congregation? Remember the *"blessed servant"* provides portions of food, at the end of the world, according to Matthew 24:44-46.

Keep in mind, most of the things we own will quickly readjust in value, in proportion to the *life-sustaining benefits they provide* to us and others. The very wealthy may experience the greatest emotional drop, if their hearts are over-attached to their money and stuff. In an extreme illustration of values shifting, it is not inconceivable to imagine how a desperate Christian mogul, with millions in assets - snared in banks, might think it a reasonable trade to give a hundred acres of freeway front property, in exchange for a hundred fish hooks to feed his family. You get the point.

In fact, if you only had one passage memorized to encourage yourself, it might be Philippians 3:7-11. Here, Paul proclaims his all-for-one, refocused earth-to-heaven revaluing: *"What things were gain to me, those I counted loss for Christ. Yea doubtless, and I count all things but loss for the excellency of the knowledge of Christ Jesus my Lord: for whom I have suffered the loss of all things, and do count them but dung, that I may win Christ, and be found in him, not having mine own righteousness, which is of the law, but that which is through the faith of Christ, the righteousness which is of God by faith: that I may know him, and the power of his resurrection, and the fellowship of his sufferings, being made conformable unto his death;* **if by any means I might attain unto the resurrection of the dead."**

Three or four families who approach a pastor, should expect direct answers to the many serious questions that arise from this worldview. Are your fellow church members steadfast in the faith to become martyrs, if that becomes necessary? You might begin leadership dialog and plan alternatives ahead of time; if 'A' happens, we will do this or that; or if 'B' occurs, we will buy X and Y. Speak words of faith daily. Faith is crucial for the Lord to show us our pathway. The greatest, remember, is a servant of all. Murmuring can cause failure of any plans. From the Bible, *"if a house be divided against itself, that house cannot stand"* (Mk 3:25). The Psalmist: *"to him that ordereth his conversation aright will I show the salvation of God"* (Psalms 50:23)

As events become obvious, church groups might choose to hold special promise ceremonies, committing to follow Christ at all costs, for encouragement. Assembling in congregations to make faith confessions, pledges, and vows can help cement decisions and bolster weaker believers. Rings and token gifts can serve to symbolize the agreements and may be worn or kept in memory of the unity and commitments. (God said to gather and pray at this time, see Joel 2:12-19). Widows and singles can bring their provisions and seek to join capable groups. Large congregations can reduce to local groups as things intensify. ***Oh Lord God, we thank You for the supernatural courage that You give us!***

Begin a Time of Prayer, Creative Dialog,: Planning and Educating Others.

Now that you've read to here, don't react; instead, may I suggest you take time alone to pray, study, and perhaps fast. This book has a lot of unique and potentially unnerving information. You might spend time alone with the Lord. Leaders, throughout the Bible, made time alone when they sensed God leading in a new direction. My hope is that constructive dialog will take place, scrutinizing this scenario, modifying, reinforcing, adding, or eliminating ideas - all filtered through Scripture. I am probably blind to things that God has made obvious to you; everyone's perception and situation is different.

A reaction to the information in this book should not include prayer-less or rash moves. Examples of diligence however, might include appraising jewelry to sell for food money, and thinning out excess stuff maybe with yard sales, Craigslist, and EBay. What things of yours are useless, if you saw signs the tribulation was starting this month? Whether to take more extreme action, like delaying mortgage payments or selling a car, may be serious options if things start happening, and food money is simply not available. Even so, overreacting or selling out too early is poor judgment and may lead to intense stress if what you are seeing is a false start, and events are delayed for years longer.

This era should not be faced alone. Join or develop a group to talk with and plan, go over this book; don't be a maverick Christian. There was talk of fencing some cities, if Y2K became chaotic. You might consider an alternate out of town location, to begin preparing and storing supplies. With food, friends will be easy to find! Introducing this information can help make alliances with other families. Use your imagination. Go through your things to categorize what is, and is not useful, or that can be turned into quick money for food and camping equipment. Sponsoring or starting a neighborhood garage-based bulk-food retail supply and equipment sales outlet would help others, and with the reasonable profits, provide for yourself a more extensive inventory. Suppliers will not lack customers, if people are informed.

Our loving Father has been forever a God of order. Ignore temptations to be afraid, quote Bible verses and sing instead. Since the joy of the Lord is our strength, we must always be laboring to enter His rest, and maintaining a focus on His sacrifice of love. Relax, truly nothing is better than going to heaven! The things that will happen in the great tribulation are actually what has always gone on here on earth, since the days of

Adam: people living and people dying. Only now it is in a merciful decision-demanding concentrated time period. As blood-bought Christians, ALL of our sins are forgiven. When Jesus died, He took the sins of the whole world away: *"Behold the Lamb of God, which taketh **away** the sin of the world"* (John 1:29). People need to hear that their sins are not counted against them - if they only call on Jesus to save them. Those of us who do end up in the ranks of waiting martyrs, must not listen to Satan's lie that it was because of our sin... Just the opposite; if we are standing there, **it is God wanting to promote our encouragement and strength!** It actually means God knew we could handle the special challenge. Resolutely following through is an opportunity to give confidence to those behind us, and finally, greater eternal rewards!

Except for the sign of Jerusalem being surrounded in Luke 21:20, it is not known if the pivotal *"abomination of desolation"* signal event will come abruptly with no prior warning, or if some publicly recognizable activity will enable us to discern, over time, that the momentous event is near, allowing us a little more time to notify others and prepare ourselves. The idea of it coming as a "snare" doesn't favor the latter (prior warning). How frustrating to have been warned early, only to face your wealth trapped in accounts, while you hold useless stacks of paper assets, have a garage full of toys, own a house full of finery with a meager pantry, and everyone you know in shocked surprise and equally unprepared. Wouldn't an ideal faith-response to this message be, to begin notifying others of these prophetic insights now, as you've learned them, while also steadily turning of all your non-necessities into food inventory, while food is cheap. Then you use the food to support as many people as possible, and finally end up (fully invested in others, aka- penniless) eating your last stored bites, as you look up to see Jesus enter your view on the horizon, or (blink) you are raptured?

Because of God's Love for Us, This Will Be Possible!

Consider, Almighty God has shown us a specific detailed warning sign for the very first day that we are to start counting off the days remaining to endure; then 1334, 1333, 1332... what mercy! By calculating and collecting a food storage inventory, according to the 1335 days of the great tribulation, groups of disciplined Christians working together could bring that food-planning ideal to reality! To determine the amount of inventory needed, count the number of people and the remaining number of days to endure. In other words, figure the total number of meals at one pound dried food/per adult/per day. Adjustments can be made, to either allow more people to enter and eat, or a method initiated for rationing remaining food. Having scales for weighing and managing food will be a must.

God never ceases to care for his people even in unusual ways. The crafty Apostle Paul survived by being let down with a rope. Reconsider a weighty denominational obligation. Be assured the Lord is not applauding mindless religious repetition of any kind. He is only interested in a relationship with you, and you responding to His Word. Shedding dry monotonous religious traditions may facilitate your down-the-rope-and-outside-the-city type responsiveness. The prostitute Rahab saved Israeli spies with an outright lie that they had already left; it ended up saving her household! Later the Apostle Paul noted in Hebrews 11:31 that she was saved by faith when *"she received the spies with peace."* Throughout history, we read of times that faith in God demanded bold action; the great tribulation era will be another such time.

Some Additional Thoughts, Observations and Cautions

Almighty God has allowed trials to prove or test people at all different points in history - often specifically with a lack of food and water.

Deuteronomy 8:2-3 tells of one such test:

> *"Thou shalt remember all the way which the*
> *LORD thy God led thee these forty years*
> *in the wilderness, to humble thee, and to*
> *prove thee, to know what was in thine*
> *heart, whether thou wouldest keep his*
> *commandments or no. And he humbled*
> *thee, and suffered thee to hunger, and fed*
> *thee with manna, which thou knewest not,*
> *neither did thy fathers know; that he might*
> *make thee know that man doth not live by*
> *bread only, but by every word that*
> *proceedeth out of the mouth of the Lord*
> *doth man live."*

— DEUTERONOMY 8:2-3

For 2000 years, generations of people have lived in a comfortable, Grace "soaked" world. Do we dare now criticize God, if He has chosen to humble and prove what is in the hearts of earth's last living people, not for forty but a mere three and a half years or so? Will people calling themselves *Christians* deny Him and fall away, or is today's modern faith, in truth, an enduring faith that pleases God? It seems the Lord will similarly test us to see. We must never murmur against God.

There are other areas of concern that should be examined. Blind allegiance to a person or a denomination could ensnare ungrounded individuals and assemblies.

Our human tendency is to over-trust spiritual leaders. Don't do this. To grant people in authority, governing control of your humble vulnerable naturally submissive spiritual life, can leave you open to be easily misguided and manipulated. History shows that wayward spiritual leaders can mislead, take advantage of and herd people like sheep, and otherwise destroy followers, particularly when using the name of God. If what any man or group promotes **goes against the plain reading of Scripture, their ideas must be subjugated or disregarded. It is probably wise to reconsider your association with them.** God's Holy Word is your and my first and final guide; it is both personal and reliable. How dangerous to rely on someone else to apply God's eternity-influencing meanings, **blindly** to your life.

The utmost extent to which we are to cling to God alone and if needed, go against even those in our own family, is shown in Deuteronomy 13:6–9. We will each stand and give account to God for our own words and actions, regardless of where our reasoning and motivation came from. On judgment day,

"He said so," will sound pitifully shallow if it conflicts with, It is written. Eternal safety will always remain tethered securely to the whole counsel of the Scriptures."

Many people today are health conscious and pursue a quality of life that includes exercise and healthy eating. The Bible teaches that bodily exercise does profit a little. However, an over-focus on ones own well being in the face of danger, could easily switch to a preserve-myself-at-any-cost mentality. Hurting innocent people, to preserve your own life, is not justified in the Bible. In fact, the Bible teaches that we are to hate our life in this world, in comparison to our valuable eternal life. We should be prepared to quietly give up our life if need be:

> *"Verily, verily I say unto you, Except a corn of wheat fall into the ground and die, it abideth alone: but if it die, it bringeth forth much fruit. He that loveth his life shall lose it; and he that hateth his life in this world shall keep it unto life eternal"*
>
> — (JOHN 12:24-25)

Our lives should be focused on the good of others:

> *"For to me to live is Christ, and to die is gain ... For I am in a strait betwixt two, having a desire to depart, and to be with Christ;*

which is far better: Nevertheless to abide in
the flesh is more needful for you"

— (PHILIPPIANS 1:21, 23-24)

If death comes, it is far better gain, while self-preservation does not justify hurting innocents.

FAITH HAS FEET

Finally dear reader, for you to have this knowledge is like being the first to see a massive hurricane moving slowly toward your loved ones and acquaintances. They have not heard of this crisis coming. Please share this new information so others will not be shocked and confused should things begin earlier rather than later. The Bible's description of the great tribulation is difficult to consider with any confidence of grasping the total impact it will be to the lives of those who face it. Not being certain of a storm's exact direction or full intensity or precise timing, does not stop wise people from preparing their homes, when they know one is coming. Yet, once again, others can only do so if warning is given them by you, this won't be on any radio or TV.

Naturally, everyone will form their own response, or non-response; that cannot be your or my concern.

What if the recipients of your early notification had time to notify others, and in heaven - from your diligence, you find many thousands had time to consider, share and wisely respond? Certainly, when the tribulation begins, it will be much more invasive and disruptive than any weekend storm. Posting copies of this book cover can alert and inform.

Having this general knowledge, before the start of events, will allow individuals a time for reviewing their doctrines in light of a lot of new information. *Most importantly, people can adjust and establish correct spiritual decisions.* As well, groups can give attention to physical and emotional readiness and unity. Informing people can allow them to stay clear of the level of confusion and fear that those caught in the prophesied "*snare,*" will most certainly experience. After notifying people, you could rest assured knowing those you care about, will have read (or for later - have it in paperback form) the clear warning God gives in Revelation, of the fiery eternal destiny of people who take the mark of the beast.

If you still do not get the purpose of this book, read Matthew Chapter 24 more closely. It culminates with verse 45 and 46, *"Who then is a faithful and wise servant, whom his lord hath made ruler over his household, to **give them meat in due season**? Blessed is that servant, whom his* lord **when he cometh shall find so doing**". We are to gather provisions for our households. Quick question, are you leaning at this point, toward an NB or SA approach?

The greater the number of people who see and understand these things, it stands to reason, the more of a fizzle the anti-Christ's fame and influence will be. Each of the prophecies will certainly be fulfilled, at some point, but how many more people can be prepared or at least not enter the tribulation utterly confused and at risk of being deceived? Can we as a body of individual believers show off to God, our unwavering zealous faith? Jesus posed an interesting question for our generation to answer: *"When the Son of man cometh, shall he find faith on the earth?"* (Luke 18:8). Sharing is not just thoughtful it is crucial for people to know about these needs early and to know how the Lord expects them to prepare to meet them. Emailing pictures of this book's covers can provoke interest enough for truth seekers to investigate.

The *"Lord's Prayer"* will certainly take on new and greater personal passion during this time: *"Thy kingdom come. Thy will be done on earth, as it is in heaven. Give us this day our daily bread" "and lead us not into temptation, but deliver us from evil"* (Matthew 6:10-13). Please pass out this information. The Bible is practical. God has pre-mapped these times for us. As His children, we love light and naturally despise living in the dark! The Apostle Paul writes, *"But ye, brethren, are not in darkness, that that day should overtake you as a thief. Ye are all the children of light, and the children of the day: we are not of the night, nor of darkness"* (1 Thessalonians 5:4-5).

Matching Evidence Warrants Careful Consideration

Some people, who began believing Jesus was the Messiah, mocked in disgust the resistant religious leaders who were standing nearby, challenging them:

> *"When Christ cometh, will he do more miracles than these which this man hath done?"*

> — (JOHN 7:31)

This is a pressing thought for us. How many prophetic signals should it take to convince us? If statistics were used to analyze the probability of these many diverse indications having occurred or that are now in plain alignment with Scripture to be fulfilled, the result would be considerably against the assembled information in this book, being a massive global coincidence. Further, we understand that our loving heavenly Father is in control of this world; is it in His heart that He would allow such a broad array of odd identifying features to harmonize into a distracting deception, *when His instruction to us repeatedly, is to be watchful for these very signs? Of course not. It might be uncomfortable but we must believe.*

The Lord has prepared the Bible for all people to read, understand and obey. In the past, He has used fishermen, common tax collectors, and a general cross section of a culture to emphasize this. God has not subjugated the world to sit outside the doors of theological scholars, breath-holding for them to interpret the Bible and share with us. The real question, like all past transitioning people, will we dare to boldly believe the clear wording and indications given by God in the Bible, with child-like faith? This book is written in sincere and simple obedience to the Lord's Revelation 13:18 command: *"Here is wisdom. Let him that hath understanding count... the number of a man."* You have effectively fulfilled this prophecy when you counted his number. So what now? It is written: *"and they shall place the abomination that maketh desolate. And such as do wickedly against the covenant shall be corrupt by flatteries: but the people that do know their God shall be strong and do exploits.* <u>*And they that understand among the people* **shall instruct many"**</u> (Daniel 11:31-33).

Should we ignore this instruction, expecting a different prince later, one whose name calculates to 666, confirmed in two languages by the very calculating system common with John of Revelation? Are we to expect this other prince to have a lineage chart showing that he is a descendant of King David? Are we to imagine another prince coming, with secretive control of the majority of this world's food distribution, while exactly matching each of the many specific and odd identifying symbols of the Bible? Will another Jewish group arise, who eagerly anticipates a "messiah" son of David? Are we to expect this other group, having as their primary mission to sacrifice on an altar of stone, and to possess such a unique stone - made without tools, while also having as their primary goal to put it on the single spot on earth that God has chosen to perpetually put His eyes, His heart, and His name? Should we expect this alternate group to replace the current people who have already proven their intent, by attempting to place the stone to begin sacrificing? Having this knowledge, what excuses will we give to the Lord, if the abomination occurs and anti-Christ begins to reign, but we abandoned his Revelation directive to allow others to count, to recognize, and be ready?

An objective court would likely view the quantity and quality of these very unique prophetic fulfilled features, as evidence beyond a reasonable doubt. I'm watching, preparing, and sharing. I've determined that, as for me and my house, like Joshua, we will serve the Lord (Joshua 24:15). The very definition and nature of "biblical faith" is action that stems from belief in the words of the Bible. Moreover, faith is actually true faith only when it results in good works; remember that *"faith without works is dead"* (James 2:26). Now, its your move. I have stretched my faith and written **Dear Christian, Do NOT Fall Away**, and this Addendum, with no visible evidence of the tribulation starting. I've written during a time in my life that is peaceful, financially stable, and filled with the love of my family and a great church. Please share the link to the book. Not everyone will purchase the book; that's okay. At least having the general information, people will have some understanding that can provide direction. If the tribulation *"snare"* becomes commonly expected, even if it begins years away, more of the tested people who are caught in it, can make godly decisions, prepare, and educate others. Knowledge of these things automatically makes you a leader!

"My People Are Destroyed for Lack of Knowledge"
(Hosea 4:6).

An encouraging biblical directive is for us to actively watch for the Lord's return. Jesus speaks of His return in Luke 21:27 and 28: *"Then shall they see the Son of man coming in a cloud with power and great glory. And when these things begin to come to pass, then look up, and lift up your heads; for your redemption draweth nigh."* Since we are told to look up as events begin, and then watch as our *"redemption draweth nigh,"* it seems that we will be able to actually observe over some period of time the coming of our redeeming Lord toward us! Again: *"Behold, he cometh with clouds; and every eye shall see him"* (Revelation 1:7). His appearing will begin *"after the tribulation of those days... then shall appear the sign of the Son of man in heaven: and then shall all the tribes of the earth mourn, and they shall see the Son of man coming in the clouds of heaven with power and great glory"* (Matthew 24:29-30).

What an overflow of contrasting emotions the great tribulation will have; the unbounded exhilaration of watching the physical return of our Lord Jesus Christ coming in the sky in like manner as He went, even while we endure rapidly collapsing surroundings. Jesus ends His sermon in Luke 21, regarding His Second Coming, with the command: *"Watch ye therefore, and pray always, that ye may be accounted worthy to escape all these things that shall come to pass, and to stand before the Son of man"* (vs. 36). Remember, a single cup of cold water given in the Lord's name receives a reward. Giving this information to people you may not even know, could answer Jesus's question, *"when the Son of man cometh, shall he find faith on the earth?"* (Luke 18:8). YES! After 2000 years, the Lord has multitudes of people around the world, who both claim they believe the Bible, and have the independent faith to recognize prophetic signs and **respond**. Again, it is written: ***"They that understand among the people <u>shall instruct many</u>"*** (Daniel 11:33).

This information is so important; yet it's hardly being addressed or even considered among believers. It's too volatile for TV, too seismic for radio. Yes, it can seem overwhelming, and yes it can take your breath away. God lives inside you and me, speaks through us, and acts through us if we are a born again member of his family. Dear Christian, stand up straight: you and I are children of the King of kings; we must all live and talk like it! God has given the world signs, like counting the number of anti-Christ's name, to provide an introduction and verification to end-time understanding. By these, He intends the whole planet to be alerted! Few people recognize when true earth-shaking moments occur in their lives. Can you see this moment, and will you participate in an informative effort?

The depth of despair that our clueless and confused relatives and the people of the world will feel, if they enter the great tribulation ignorant, would be beyond comprehension! Such confusion could lead to ensnarement by a satanic, deceptive system. It takes faith when things are peaceful to make preparations and distribute this information. Please do not let doubt or fear prevent you from boldly helping as many people as possible, to at least be aware of these things. The tribulation will come as a snare upon us all. God has been and is forever merciful: *"Except that the Lord had shortened those days, no flesh should be saved: but for the elect's sake, whom he hath chosen, he hath shortened the days"* (Mark 13:20).

As This Book Is Shared, People Will Have The Opportunity To:

- See the Lord's love in history, the warning signs, prior to judgment!
- Know that someone cares for them and is praying for their endurance!
- Trash conflicting beliefs of an ultra-ancient aged earth, mythical monkey men, and an incoherent bumbling Creator!
- Embrace an independent faith, cleansed from religious ritual and scientific deception!

- See photo evidence the Lord spoke accurately to our enslaved forefathers to free them, and trust that He is speaking truth to us now!
- Gain confidence in the words of the Bible from the very first verses, and pray for his or her simple salvation!
- Clarify, Biblical fact from fiction, when the rapture may occur!
- Accept and Value the unlimited power of God, to lead them daily!
- Trust their Creator's love even in chaos, to lead us to His eternal home!
- Obey God's compassionate prophetic command to "*let*" others with "*understanding count the number of a man*!"
- Share this knowledge with others, and rest, knowing friends and family will not be clueless in catastrophe and calamity!
- Make crucial decisions wisely from a grounded faith, rooted in a comprehensive worldview!
- Know to estimate, gather, and give their households portions of meat, to endure the fixed written tribulation time period!
- Have an opportunity to study and apply this detailed Addendum of ideas and preparation insights!

- Face chaos into martyrdom if need be, knowing all creation, particularly man's prophesied calamitous closure, validates their Biblical faith!

Every generation tends to believe that they will not fail God, as have those previously; it is our human nature. Pride and hubris however, take no one far with God. Proverbs 16:18, Faith, Hope, and Love make up our spiritual personality now and in the future. Only in heaven will we know the final help this information was to others. God is our Father, and Jesus is our Lord and Savior, we have the Holy Spirit living inside of us – the same power that raised Christ from the dead! You and I will remain forever in God's family, among God's chosen! Rejoice! Heaven is now close at hand!

ENCOURAGING BIBLE VERSES: FOR DAILY READING

It Is Written:

"He will swallow up death in victory; and the Lord God will wipe away tears from off all faces; and the rebuke of his people shall he take away from off all the earth: for the Lord hath spoken it."

— (ISAIAH 25:8)

"The fear of man bringeth a snare: but whoso putteth his trust in the Lord shall be safe"

— (PROVERBS 29:25).

"Thou shalt fear the Lord thy God, and serve him, and shalt swear by his name"

— (DEUTERONOMY 6:13).

"Alas! For that day is great, so that none is like it: it is even the time of Jacob's trouble; but he shall be saved out of it"

— (JEREMIAH 30:7).

"Great is my boldness of speech toward you, great is my glorying of you: I am filled with comfort, I am exceeding joyful in all our tribulation"

— (II COR. 7:4)

"In all things approving ourselves as the ministers of God, in much patience, in afflictions, in necessities, in distresses, in stripes, in imprisonments, in tumults, in labours, in watchings, in fastings; by pureness, by knowledge, by longsuffering, by kindness, by the Holy Ghost, by love unfeigned, by the word of truth, by the power of God, by the armour of righteousness on the right hand and on the left, By honour and dishonour, by evil report and good report: as deceivers, and yet true;

as unknown, and yet well known; as dying, and, behold,
we live; as chastened, and not killed; as sorrowful, yet
always rejoicing; as poor, yet making many rich; as
having nothing, and yet possessing all things"

— II COR. 6:4-10

"Be sober, be vigilant; because your adversary the devil, as
a roaring lion, walketh about, seeking whom he may
devour: whom resist steadfast in the faith, knowing that
the same afflictions are accomplished in your brethren
that are in the world. But the God of all grace, who
hath called us unto his eternal glory by Christ Jesus,
after that ye have suffered a while, make you perfect,
stablish, strengthen, settle you. To him be glory and
dominion for ever and ever. Amen"

— (I PETER 5:8-11)

"For behold, the day cometh, that shall burn as an oven;
and all the proud, yea, and all that do wickedly, shall
be stubble: and the day that cometh shall burn them up,
saith the LORD of hosts, that it shall leave them neither
root nor branch. But unto you that fear my name shall
the Sun of righteousness arise with healing in his
wings; and ye shall go forth, and grow up as calves of

the stall. And ye shall tread down the wicked; for they

shall be ashes under the soles of your feet in the day

that I shall do this, saith the LORD of hosts. Remember

ye the law of Moses my servant, which I commanded

unto him in Horeb for all Israel, with the statutes and

judgments. Behold, I will send you Elijah the prophet

before the coming of the great and dreadful day of the

LORD: and he shall turn the heart of the fathers to the

children, and the heart of the children to their fathers,

lest I come and smite the earth with a curse"

— (MALACHI 4)

"I will praise thee, O LORD, with my whole heart; I will

shew forth all thy marvelous works. I will be glad and

rejoice in thee: I will sing praise to thy name, O thou

most High. When mine enemies are turned back, they

shall fall and perish at thy presence. For thou hast

maintained my right and my cause; thou satest in the

throne judging right. Thou hast rebuked the heathen,

thou hast destroyed the wicked, thou hast put out their

name for ever and ever. O thou enemy, destructions are

come to a perpetual end: and thou hast destroyed

cities; their memorial is perished with them. But the

LORD shall endure for ever: he hath prepared his

throne for judgment. And he shall minister judgment to

the people in uprightness. The LORD also will be a refuge for the oppressed, a refuge in times of trouble. And they that know thy name will put their trust in thee: for thou, LORD, hast not forsaken them that seek thee. Sing praises to the LORD, which dwelleth in Zion: declare among the people his doings. When he maketh inquisition for blood, he remembereth them: he forgetteth not the cry of the humble. Have mercy upon me, O Lord; consider my trouble which I suffer of them that hate me, thou that liftest me up from the gates of death: That I may shew forth all thy praise in the gates of the daughter of Zion: I will rejoice in thy salvation. The heathen are sunk down in the pit that they made: in the net which they hid is their own foot taken. The LORD is known by the judgment which he executeth: the wicked is snared in the work of his own hands. Higgaion. Selah. The wicked shall be turned into hell, and all the nations that forget God. For the needy shall not always be forgotten: the expectation of the poor shall not perish for ever. Arise O LORD; let not man prevail: let the heathen be judged in thy sight. Put them in fear. O LORD: that the nations may know themselves to be but men. Selah

— (PSALM 9)

"Thou shalt not be afraid for the terror by night; nor for the arrow that flieth by day; nor for the pestilence that walketh in darkness; nor for the destruction that wasteth at noonday. A thousand shall fall at thy side, and ten thousand at thy right hand; but it shall not come nigh thee. Only with thine eyes shalt thou behold and see the reward of the wicked. Because thou hast made the LORD, which is my refuge, even the most High, thy habitation; there shall no evil befall thee, neither shall any plague come nigh thy dwelling. For he shall give his angels charge over thee, to keep thee in all thy ways. They shall bear thee up in their hands, lest thou dash thy foot against a stone. Thou shalt tread upon the lion and adder: the young lion and the dragon shalt thou trample under feet. Because he hath set his love upon me, therefore will I deliver him: I will set him on high, because he hath known my name. He shall call upon me, and I will answer him: I will be with him in trouble; I will deliver him, and honour him. With long life will I satisfy him, and shew him my salvation"

— (PSALM 91:5-16)

"Behold, we count them happy which endure. Ye have heard of the patience of Job, and have seen the end of the Lord: that the Lord is very pitiful, and of tender mercy"

— (JAMES 5:11)

"Not that I speak in respect of want: for I have learned, in whatsoever state I am, therewith to be content. I know both how to be abased, and I know how to abound: every where and in all things I am instructed both to be full and to be hungry, both to abound and to suffer need. I can do all things through Christ which strengtheneth me"

— (PHILIPPIANS 4:11-13)

"God is our refuge and strength, a very present help in trouble. Therefore will not we fear, though the earth be removed, and though the mountains be carried into the midst of the sea; though the waters thereof roar and be troubled, though the mountains shake with the swelling thereof. Selah. There is a river, the streams whereof shall make glad the city of God, the holy place of the tabernacles of the most High. God is in the midst of her; she shall not be moved: God shall help her, and that right early. The heathen raged, the kingdoms were moved: he uttered his voice, the earth melted. The LORD of hosts is with us; the God of Jacob is our refuge. Selah. Come, behold the works of the LORD, what desolations he hath made in the earth. He maketh wars to cease unto the end of the earth; he breaketh the bow, and cutteth the spear in sunder; he burneth the

chariot in the fire. Be still, and know that I am God; I
will be exalted among the heathen, I will be exalted in
the earth. The LORD of hosts is with us; the God of
Jacob is our refuge. Selah"

— (PSALM 46)

"Behold, happy is the man whom God correcteth: therefore
despise not thou the chastening of the Almighty: for he
maketh sore, and bindeth up: he woundeth, and his
hands make whole. He shall deliver thee in six troubles:
yea, in seven there shall no evil touch thee. In famine
he shall redeem thee from death: and in war from the
power of the sword. Thou shalt be hid from the scourge
of the tongue: neither shalt thou be afraid of destruc-
tion when it cometh. At destruction and famine shalt
thou laugh: neither shalt thou be afraid of the beasts of
the earth"

— (JOB 5:17-22)

"None of us liveth to himself, and no man dieth to himself.
For whether we live, we live unto the Lord; and
whether we die, we die unto the Lord: whether we live
therefore, or die, we are the Lords. For to this end
Christ both died, and rose, and revived, that he might

be Lord both of the dead and the living. But why dost

thou judge thy brother? Or why dost thou set at nought

thy brother? For we shall all stand before the judgment

seat of Christ. For it is written, As I live, saith the Lord,

every knee shall bow to me, and every tongue shall

confess to God. So then every one of us shall give an

account of himself to God"

— (ROMANS 14:7-12)

"Beloved, think it not strange concerning the fiery trial

which is to try you, as though some strange thing

happened unto you: but rejoice, inasmuch as ye are

partakers of Christs sufferings; that, when his glory

shall be revealed, ye may be glad also with exceeding

joy. If ye be reproached for the name of Christ, happy

are ye; for the spirit of glory and of God resteth upon

you: on their part he is evil spoken of, but on your part

he is glorified. But let none of you suffer as a murderer,

or as a thief, or as an evildoer, or as a busybody in

other mens matters.

Yet if any man suffer as a Christian, let him not be ashamed;

but let him glorify God on this behalf. For the time is

come that judgment must begin at the house of God:

and if it first begin at us, what shall the end of them

that obey not the gospel of God? And if the righteous

scarcely be saved, where shall the ungodly and the
sinner appear? Wherefore let them that suffer
according to the will of God commit the keeping of
their souls to him in well doing, as unto a faithful
Creator"

— (I PETER 4:12-19)

"God hath not appointed us to wrath, but to obtain salvation
by our Lord Jesus Christ, who died for us, that, whether
we wake or sleep we should live together with him"

— (I THESSALONIANS 5:9-10)

"For to me to live is Christ, and to die is gain"

— (PHILIPPIANS 1:21)

For this God is our God for ever and ever: he will be our
guide even unto death"

— (PSALM 48:14)

"Keep yourselves in the love of God, looking for the mercy
of our Lord Jesus Christ unto eternal life. And of some
have compassion, making a difference: and others save

with fear; pulling them out of the fire; hating even the garment spotted by the flesh. Now unto him that is able to keep you from falling, and to present you faultless before the presence of his glory with exceeding joy, to the only wise God our Savior, be glory and majesty, dominion and power, both now and ever. Amen"

— (JUDE 1:21-25)

"I saw a new heaven and a new earth: for the first heaven and the first earth were passed away; and there was no more sea. And I John saw the holy city, the new Jerusalem, coming down from God out of heaven, prepared as a bride adorned for her husband. And I heard a great voice out of heaven saying, Behold, the tabernacle of God is with men, and he will dwell with them, and they shall be his people, and God himself shall be with them, and be their God. And God shall wipe away all tears from their eyes; and there shall be no more death, neither sorrow, nor crying, neither shall there be any more pain: for the former things are passed away. And he that sat upon the throne said, Behold, I make all things new. And he said unto me, Write: for their words are true and faithful. And he said unto me, It is done. I am Alpha and Omega, the beginning and the end. I will give unto him that is athirst of

*the fountain of the water of life freely. He that over-
cometh shall inherit all things; and I will be his God,
and he shall be my son. But the fearful, and unbeliev-
ing, and the abominable, and murderers, and whore-
mongers, and sorcerers, and idolaters, and all liars,
shall have their part in the lake which burneth with fire
and brimstone: which is the second death" (Revelation
21:1-8). "Behold, I come quickly; and my reward is
with me, to give every man according as his work shall
be. I am the Alpha and Omega, the beginning and the
end, the first and the last"*

— (REVELATION 22:12-13)

*"Blessed be the God and Father of our Lord Jesus Christ,
which according to his abundant mercy hath begotten
us again unto a lively hope by the resurrection of Jesus
Christ from the dead, to an inheritance incorruptible,
and undefiled, and that fadeth not away, reserved in
heaven for you, who are kept by the power of God
through faith unto salvation ready to be revealed in the
last time. Wherein ye greatly rejoice, though now for a
season, if need be, ye are in heaviness through manifold
temptations: that the trial of your faith, being much
more precious than of gold that perisheth, though it be
tried with fire, might be found unto praise and honour
and glory at the appearing of Jesus Christ: Whom*

having not seen, ye love; in whom, though now ye see him not, yet believing, ye rejoice with joy unspeakable and full of glory: Receiving the end of your faith, even the salvation of your souls"

— (I PT 1:3-9)

"The Lord knoweth how to deliver the godly out of temptations, and to reserve the unjust unto the day of judgment to be punished"

— (II PETER 2:9)

"Now the just shall live by faith: but if any man draw back, my soul shall have no pleasure in him. But we are not of them who draw back unto perdition; but of them that believe to the saving of the soul"

— (HEBREWS 10:38-39)

"He which testifieth these things saith, Surely I come quickly. Amen. Even so, come, Lord Jesus. The Grace of our Lord Jesus Christ be with you all. Amen"

— (REVELATION 22:20-21)

May "the LORD bless thee, and keep thee: the LORD make

his face shine upon thee, and be gracious unto thee: the

LORD lift up his countenance upon thee, and give thee

peace"

— (NUMBERS 6:24-26)

See also, I Peter, chapters 4 and 5; II Corinthians, chapters 4-6 and 7:1-4; Revelation 2:10,11; Luke 14:26; Philippians 1:20; Psalms 32:7-8; 9:7-11.

NOTES

FOREWORD

1. Blaise Pascal Quotes - BrainyQuote

1. HUMAN ORIGIN:

1. what-age-accept-christ-statistic.jpg (637×414) (ministry-to-children.com)
2. Barna http://crossexamined.org/youth-exodus-problem/
3. Make the lie big, make it simple, keep saying it, and eventually they will believe it. ~ Adolf Hitler. Is Modi's 'Acche Din' one of the above kind? - Quora
4. https://www.brainyquote.com/quotes/adolf_hitler_382802
5. www.ICR.org
6. ICR Impact, Article No. 346, Institute for Creation Research, April 2002
7. ICR Impact, IBID
8. Ham, Ken. Answers In Genesis. Creation video series,
9. www.changingworldtech.com
10. www.Creationworldview.org
11. Weiland, C., "Earth: How Old Does It Look?" Creation Magazine 23 (1) Dec., 2000- Feb., 2001,
    p. 10
12. "Trends in World Population. Encyclopedia Britannica. 2000 CD.
13. Article: "Where Are All the People"; Creation Magazine, vol. 23, No.3, June - August 2001
14. www.answersingenesis.org/Docs/4232cen_s1997.asp
15. Gish, D.T. Evolution: The Challenge of the Fossil Record. El Cajon, California: Creation Life Publishers, 1985, p. 121. (Note:

book provides extensive research of the alleged hominid ancestry, debunking alleged 'monkey-man' skulls.)

16. Darwin, Charles. Origin of the Species. Chapter VI.

17. www.arkdiscovery.com

18. Bible Questions Answered | GotQuestions.org (700,000 Bible answers)

19. Cuozzo, JACK. Buried Alive, The Startling Truth About Neanderthal Man. Master Books Inc., 1998

20. http://www.whychurch.org.uk/trends.php

2. THE ULTIMATE FINISH LINE

1. Mott, Jeff. The Rapture When? Norman, Oklahoma: Lion and Lamb Ministries. Additional copies of Jeff Mott's, "THE RAPTURE WHEN?" article are available for $2.00 each from: Lion and Lamb Ministries, P.O. Box 720968, Norman, Oklahoma 73070.

4. GOD'S JUDGMENT, TIMING, AND THE BIBLICAL SIGNS

1. Www.templemountfaithful.org

2. www.inner.org/gematria/gematria.htm

3. Cohen, Tim. The Antichrist and a Cup of Tea. Aurora, Colorado: Prophecy House, Inc., 1998, p. 181 (This book is well documented; with 444 pages. It substantiates the broad yet secretive control "prince" Charles is permitted by God to have in this world. For those who will pursue personal verification, this book is a good place to start.)

4. Holden, Anthony. "prince" Charles. (New York: Athenaeum, 1979), xxii. Royalty, 1994, Vol. 13 no. 2, p. 176.

5. Cohen, Ibid, p. 200. A picture showing the queen placing the coronet cap on Charles head, with a prominent view of the engraved dragon centrally positioned on the back of the Queen's throne, and Prince Phillip seated to the side.

6. Holden, Ibid, pp. 191-192.
7. Cohen, Op. Cit, p. 15.
8. News Brief from London, Fox News Life, March 7, 2002
9. Webster's Dictionary. 1913
10. Cambridge Advanced Learners Dictionary
11. The Telegraph, December 2011

5. WHAT ON EARTH SHOULD BE OUR RESPONSE?

1. www.idpconsultinggroup.com, Normalcy Bias Vs. Situation Awareness

www.ingramcontent.com/pod-product-compliance
Lightning Source LLC
Chambersburg PA
CBHW070617220526
45466CB00001B/31